RAND COUNTERINSURGENCY STUDY • VOLUME 2

Counterinsurgency in Iraq (2003–2006)

Bruce R. Pirnie, Edward O'Connell

WITHDRAWN
UTSA LIBRARIES

Prepared for the Office of the Secretary of Defense

Approved for public release; distribution unlimited

 NATIONAL DEFENSE RESEARCH INSTITUTE

The research described in this report was prepared for the Office of the Secretary of Defense (OSD). The research was conducted in the RAND National Defense Research Institute, a federally funded research and development center sponsored by the OSD, the Joint Staff, the Unified Combatant Commands, the Department of the Navy, the Marine Corps, the defense agencies, and the defense Intelligence Community under Contract W74V8H-06-C-0002.

Library of Congress Cataloging-in-Publication Data is available for this publication.

ISBN 978-0-8330-4297-2

Library
University of Texas
at San Antonio

Cover Image Courtesy of 4th PSYOP Group, Ft. Bragg, NC

The RAND Corporation is a nonprofit research organization providing objective analysis and effective solutions that address the challenges facing the public and private sectors around the world. RAND's publications do not necessarily reflect the opinions of its research clients and sponsors.

RAND® is a registered trademark.

Cover Design by Stephen Bloodsworth

© Copyright 2008 RAND Corporation

All rights reserved. No part of this book may be reproduced in any form by any electronic or mechanical means (including photocopying, recording, or information storage and retrieval) without permission in writing from RAND.

Published 2008 by the RAND Corporation
1776 Main Street, P.O. Box 2138, Santa Monica, CA 90407-2138
1200 South Hayes Street, Arlington, VA 22202-5050
4570 Fifth Avenue, Suite 600, Pittsburgh, PA 15213-2665
RAND URL: http://www.rand.org/
To order RAND documents or to obtain additional information, contact
Distribution Services: Telephone: (310) 451-7002;
Fax: (310) 451-6915; Email: order@rand.org

Preface

This monograph is one of a series produced as part of the RAND Corporation's research project for the U.S. Department of Defense on how to improve U.S. counterinsurgency (COIN) capabilities. It should be of interest to persons in the government who are concerned with COIN issues and to scholars working in this field. The project will culminate in a report that builds on these earlier efforts.

As of spring 2007, when field research for this monograph was completed, Iraq appeared to have slid from insurgency into civil war. The U.S. failure to focus on the protection of the Iraqi population in the preceding four years had contributed significantly to the subsequent increase in insurgency and sectarian violence. In the security vacuum that ensued, Iraqi citizens were forced to engage in a Faustian bargain—often looking to bad actors for protection—in order to survive. The failure of the United States and the Iraqi government to subdue the Sunni insurgency and prevent terrorist acts—punctuated by the 2006 destruction of the Golden Dome—produced an escalation of Shi'a militancy and sectarian killing by both militia and police death squads.

However, by early 2008, when this monograph was published, the security situation in Iraq had started to stabilize due to a number of factors: a Sunni reaction to al Qaeda excesses, a pullback of Shi'a militias from anti-Sunni violence and confrontation with coalition forces, a decrease in externally supplied armaments, and strides made by U.S. and Iraqi forces in gaining control of important areas in the country, including western provinces such as Anbar and parts of Baghdad. It should be noted that as of this publication date, it is still not clear how the political-security situation in Iraq will eventually turn out. In particular, the authors maintain considerable doubt as to whether Iraq can reconcile the divisions between the Sunni, Shi'a, and Kurdish elements of the population. Nevertheless, the reduced level of violence as of early 2008 was an encouraging development.

That said, the authors' view is that our examination of U.S. political and military challenges in Iraq from 2003–2006 has important implications for improving future counterinsurgency strategy and capabilities. Iraq presents an example of a local political power struggle overlaid with sectarian violence and fueled by fanatical foreign jihadists and persistent criminal opportunists—some combination of forces likely to be replicated in insurgencies in other troubled states in the future. In that sense, this monograph highlights national capability gaps which persist despite the adoption and improved execution of counterinsurgency methods in Iraq.

This analysis was sponsored by the U.S. Department of Defense and conducted within the International Security and Defense Policy Center of the RAND National Defense Research Institute, a federally funded research and development center sponsored by the Office of the Secretary of Defense, the Joint Staff, the Unified Combatant Commands, the Department of the Navy, the Marine Corps, the defense agencies, and the defense Intelligence Community.

For more information on RAND's International Security and Defense Policy Center, contact the director, James Dobbins. He can be reached by email at James_Dobbins@rand.org; by phone at 703-413-1100, extension 5134; or by mail at the RAND Corporation, 1200 South Hayes Street, Arlington, Virginia 22202-5050. More information about RAND is available at www.rand.org.

Contents

Preface iii
Figures ix
Tables xi
Summary xiii
Acknowledgments xxiii
Abbreviations xxv

CHAPTER ONE
Overview of the Conflict in Iraq 1
The Ba'athist Regime 2
The Invasion of Iraq 5
The Occupation of Iraq 9
The First Priority: Setting Up a Constitutional Government 10
The Spiral Downward Begins (Spring 2004) 11
Benchmark One: Holding Iraqi Elections 13
Islamic Extremist and Sectarian Violence Begin 15
A U.S. Approach Hesitantly Unfolds 17

CHAPTER TWO
Armed Groups in Iraq 21
Overview 22
Kurdish Separatists 24
Sunni Arab Insurgents 25
Violent Extremists 28
Shi'ite Arab Militias 31

Criminal Gangs 32
Insurgent Use of Terrorism 32

CHAPTER THREE
Counterinsurgency in Iraq... 35
Organization and Recognition of the U.S. COIN Effort Is Slow
  to Unfold 35
Traditional U.S. Military Forces May Need to Be Adjusted 36
  Fallujah 38
  Tal Afar 39
  Baghdad 41
  Air Support 43
  Combating Improvised Explosive Devices 44
  Detainee Operations 46
U.S. Development and Support of Iraqi Forces 49
  The Iraqi Police 49
  The Iraqi Armed Forces 51
Assessing Progress in Counterinsurgency 52
  Iraqi Casualties and Displacement 52
  The Iraqi Economy 53
  Iraqi Opinion 58

CHAPTER FOUR
Accounting for Success and Failure 61
Understanding Iraqi Society 61
Little Planning for the Occupation of Iraq 62
The Impact of a Lack of International Support for the War 63
The Disastrous Effects of Prematurely Dismantling the Ba'athist
  Regime 64
The Challenge of Building a New Iraqi State from Scratch 65
Instituting a New System of Justice 66
Undertaking the Reconstruction of Iraq 67
The Consequences of Failing to Maintain Security Early On 69
  Military Missions 69
  Lack of Infiltration and Tips Hinders Intelligence on the Insurgency ... 71

CHAPTER FIVE
Building Effective Capabilities for Counterinsurgency 73
Use of Force 75
Public Safety and Security 78
Partnering with and Enabling Indigenous Forces 80
Reporting on the Enemy and Infiltration 83
Provision of Essential Services 86
Informing and Influencing Operations 88
Rigorous and Coordinated Detainee Operations 89

CHAPTER SIX
Recommendations 91
Development of Strategy 91
Coalition-Building 92
Planning Process 92
Unity of Effort 93
Interagency Process 93
Host-Nation Governance 94
Funding Mechanisms 94
Counterinsurgency as a Mission 94
Protection of the Indigenous Population 95
Personnel Policy 95
U.S. Army Special Forces 95
Partnership with Indigenous Forces 96
Policing Functions 96
Brigade Organization 97
Gunship-Like Capability 97
Intelligence Collection and Sharing 97

Bibliography 99

Figures

S.1.	Illustrative Lines of Operation	xvii
S.2.	Iraqi Civilian and Police Deaths, by Cause	xix
3.1.	Internal Displacement in Iraq in 2006	54
3.2.	Iraqi Civilian and Police Deaths, by Cause	55
5.1.	Illustrative Lines of Operation	74

Tables

3.1.	Iraqi Reconstruction Through December 2005	55
3.2.	Degrees of Iraqi Support for Forces in al Anbar Province in 2006	59
3.3.	Iraqi Trust of Information Sources	60
5.1.	Use of Military Force and Some Implied Tasks	79
5.2.	Security of the Population and Some Implied Tasks	81
5.3.	Enabling Indigenous Forces and Some Implied Tasks	82
5.4.	Intelligence on the Enemy and Some Implied Tasks	85
5.5.	Essential Services and Reconstruction and Some Implied Tasks	87
5.6.	Information Operations and Some Implied Tasks	88
5.7.	Detainee Operations and Some Implied Tasks	90

Summary

Background

The United States is in the fifth year of trying to combat an insurgency that began when it invaded and occupied Iraq. The conflict in Iraq involves a mixture of armed groups whose motivations vary, but three of these groups are united at the *transactional* level by a simple, common theme: The occupation of Iraq by foreign forces is bad. Some insurgents are fighting for political power inside post-Saddam Iraq. Others are motivated by sectarian (e.g., Sunni versus Shia) agendas. A particularly violent minority see the struggle as part of a larger global jihad, or religious struggle, against what they perceive as the strategic encroachment of the enemies of Islam. However, for a fourth group, criminals and/or opportunists, the war has been anything but "bad": It has immensely increased their prospects—if, for most, only in the short term.

Americans tend to see the Vietnam War as an analog to the Iraq War. However, while Vietnam had internal divisions, those divisions did not appear to be as fierce as those among the Sunni Arabs, Shi'ite Arabs, and Kurds in Iraq. Insurgents in South Vietnam were supported by North Vietnam and were eventually supplanted by regular North Vietnamese forces. Ultimately, the war ended with an invasion by North Vietnam. Despite these fundamental differences, U.S. forces might at least have profited from the experience in counterinsurgency (COIN) gained from fighting the Viet Cong, but this experience was largely forgotten, except by the Army's Special Forces units. Though the Vietnam and Iraq experiences are different on the surface, an unfortunate similarity between them is the difficulty the United States has in recognizing the nature of the problem and developing an effective political-military-economic solution, choosing instead to resort to technology for an effort that requires closely synchronized operational art and innovative strategies. The U.S. failure to contain the rising level of disorder in Iraq, as well as subsequent policy and military mistakes, helped create the environment in which an insurgency took hold and a civil war unfolded.

Although insurgency remains a fundamental problem, the conflict in Iraq is more complicated than simply a revolt against the Iraqi government. That government is so ineffective that the conflict more nearly resembles a many-sided struggle for power amid the ruins of the Ba'athist state. Broadly speaking, three major groups are involved at the core of the insurgency: Sunni Arabs, who have long dominated Iraq and will not accept an inferior position; Shi'ite Arabs, who are trying to assert a new primacy; and Kurds, whose primary allegiance appears to be to a new Kurdistan. Sunni Arabs are organized along a complex array of neighborhood affiliation, armed groups, tribes, and families, depending on the locale. Shi'ite Arabs are split into several competing factions with different agendas. Though the Kurds appear to be the most unified group, even they are split into two parties that fought each other in the recent past. Were insurgency the only challenge, U.S. and Iraqi government forces might at least contain the violence, but the multiple challenges of separatists, insurgents, extremists, militias, and criminals threaten to destroy the country at any moment. Violence in Iraq currently involves all of these elements:

- **Separatists and sectarianism.** Separatism and sectarianism compound Iraq's problems and appear to be increasing. For the most part, Kurds do not regard themselves as Iraqis first; they stay within Iraq as a matter of convenience and to wield political influence. Their leadership shows a wavering commitment to a unified pluralistic government. The Supreme Council for Islamic Revolution in Iraq (SCIRI) publicly advocates autonomous regions and envisions a large Shi'ite-dominated region in southern Iraq. In contrast, Sunni leadership has little interest in creating an autonomous Sunni region, if only because that region would not contain lucrative and well-maintained oil fields. However, the extremist Mujahideen Shura Council in Iraq has notionally announced the creation of a new Islamic state encompassing Sunni-inhabited areas.

- **Insurgents.** The insurgency springs from a sectarian and ethnic divide, i.e., Sunni Arab opposition to an Iraqi government dominated by Shi'ite Arabs and Kurds and a manifestation of opposition to U.S. forces. Countering this insurgency is the most urgent mission in Iraq, because success would allow the Iraqi government to concentrate on other serious non–security-related problems. To succeed, the Iraqi government must be perceived as impartial and able to protect all of its citizens. Creating such a perception is extremely difficult amid escalating sectarian violence, especially when government ministries are involved with sectarian militias and the government's partner is a foreign occupier that has largely resisted persistent pleas to protect the local population.
- **Violent extremists.** Extremists gravitate to the conflict for various reasons. Insurgency fits into a vision of protecting Muslim countries against foreign domination. On a personal level, it offers an outlet for resentment and a chance to attain personal redemption. In addition, many terrorist leaders are Salafist (fundamentalist) Sunnis, who deliberately incite sectarian violence by attacking Shi'ite civilians, attempting to justify their existence as shock troops, propagandists, and, in the early going, "organizers" supplementing local insurgent forces.
- **Shi'ite Arab militias.** Many Shi'ite Arabs depend upon militias more than upon Iraqi government forces for security. The militia leaders exert strong influence within the government, which refuses to curb their activities. The Badr Organization was created during the Iran-Iraq War, while the much larger Mahdi Army emerged during the U.S. occupation. U.S. security training missions focused on Iraqi Army development at the expense of monitoring growing infiltration of Iraqi police forces by these militias. The Mahdi Army combines security functions with social services, much like Hezbollah in Lebanon, thus becoming a quasi-

state within a state. It appears to be linked with Shi'ite death squads that abduct, torture, and kill Sunni Arabs in the Baghdad area.

- **Criminals.** Criminality continues to plague the country, and criminals hire out their services to enemies of the Iraqi government. Although few crime statistics are kept, it appears that many Iraqis consider criminality to be the greatest threat in their daily lives. The lack of organic tools and mechanisms among its chief partner, the U.S. military, to combat crime and the Iraqi government's inability to do so diminish the Iraqi government's legitimacy and its appeal for allegiance among its constituents—the Iraqi people.

The Goal and Art of Counterinsurgency

The primary goal of COIN is to protect the population in order to obtain its tacit and active support in putting down the insurgency and thereby gain its allegiance. Until recently, this key tenet of COIN has been overlooked in Iraq. Until early 2007, the U.S. COIN effort in Iraq neglected the protection of the people, a policy oversight that adversely affected the overall effort to rebuild the nation. Until and unless there are sustained and meaningful signs of will and commitment on the part of the counterinsurgents, the allegiance of a besieged populace to a government they are somewhat detached from will remain problematic. Signs of increasing allegiance would include willingness to risk providing information on insurgents, participation in civic life, holding public office, serving as police, and fighting as soldiers.

The art of COIN is achieving synergy and balance among various *simultaneous* civilian and military efforts or lines of operation (see Figure S.1) and continually reassessing the *right* indicators—not just those that are politically expedient—to determine whether current strategies are adequate. The need to continually reassess COIN strategy and tactics implies that military and civilian leaders must be willing and able to *fearlessly* and *thoroughly* call policies and practices that are not working to the attention of senior decisionmakers.

Figure S.1
Illustrative Lines of Operation

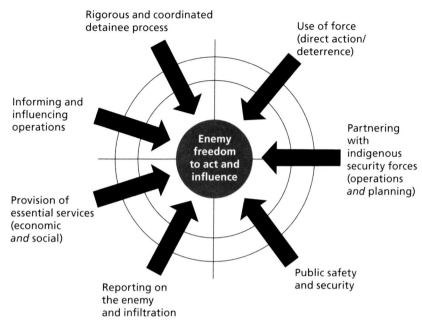

RAND MG595/3-S.1

Because COIN requires the harmonious use of civilian and military means, unity of effort is the *sine qua non* for success. Unity of effort implies that all relevant entities, military and civilian, are subject to common control in pursuit of the same strategy.

Recommendations

The United States needs to improve its ability to develop strategy and to modify it as events unfold. Strategy implies a vision of how to attain high-level policy objectives employing U.S. resources and those of its allies. It also implies reflection upon strategies that adversaries might develop and how to counter them—*counteranalysis*. Strategy should be developed at the highest level of government, by the President, his closest advisors, and his Cabinet officials, with advice from the Director of

National Intelligence and regional experts, the Chairman of the Joint Chiefs of Staff, and unified commanders. The unified commanders should link the strategic level of war with the operational level at which campaign planning is accomplished.

To successfully prosecute COIN in Iraq, the United States needs a comprehensive strategy, including a framework that carefully addresses and assesses various lines of operations and considers tradeoffs between the effects emanating from them. The efforts required for success are mutually reinforcing, implying that they must all be made simultaneously, though the appropriate weight of effort may vary over time and location. Counterinsurgency is a political-military effort that requires both good governance and military action. It follows that the entire U.S. government should conduct that effort. The following recommendations would assist it in developing capabilities to conduct COIN:

- Focus on security of the population as the critical measure of effectiveness. For too long, this was not a priority in Iraq. Exceptional efforts must be taken to remove primary threats to the civilian populace. In Iraq, senior military commanders focused almost all efforts on roadside bombs and their impact on U.S. forces, rather than the suicide-bomber problem and its terrible impact on the safety of civilians, which became increasingly evident in the summer of 2004 (see Figure S.2).
- Allow Army Special Forces to focus on training and operating together with their indigenous counterparts. Command arrangements should assure that Special Forces harmonize with the overall effort, while allowing scope for initiative. In addition, the Army should conduct training and exercises prior to deployment to educate conventional-force commanders in special operations, especially those involving *unconventional* warfare—practiced surprisingly little in Iraq.
- Develop a planning process that embraces all departments of the U.S. government and is on the same battle rhythm as troops in the field. In the context of a national strategy, an office with directive authority should assign responsibilities to the various departments, assess their plans to discharge these responsibilities, request

Figure S.2
Iraqi Civilian and Police Deaths, by Cause

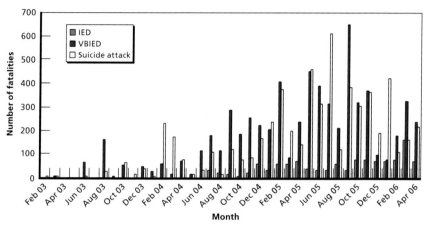

SOURCE: Data compiled by merging IraqBodyCount.org database and RAND-Memorial Institute for the Prevention of Terrorism (MIPT) Terrorism Incident database. As of October 1, 2007: http://www.iraqbodycount.org and http://www.tkb.org.
NOTE: IED = improvised explosive device; VBIED = vehicle-borne improvised explosive device.
RAND MG595/3-S.2

changes as appropriate, and promulgate a political-military plan. This plan should have enough operational detail to serve as an initial basis for execution of a COIN campaign.
- Quickly develop a coherent and balanced COIN strategy. In Iraq, the United States did not have a clear COIN strategy or plan for more than three years. Senior military commanders and planners must establish an adequate mechanism with which to constantly assess performance in COIN operations. Senior military commanders must adapt/adjust/modify strategy and tactics to meet the ever-changing demands of those operations. Commanders must closely monitor changing trends on the battlefield. In Iraq, senior military commanders have been slow to understand and adapt to the change in the enemy's strategy and tactics.
- Assure unity of effort at the country level and provincial levels, encompassing all activities of the U.S. government, civilian and military. At the country level, there should be one individual with

authority to direct all aspects of the U.S. effort. In Iraq, interagency tensions have hampered the COIN effort.
- Put the interagency process in Washington on a wartime footing to conduct any COIN operations requiring large-scale U.S. forces. This process should support the person appointed by the President to prosecute the campaign within the parameters of the national strategy. The process should be structured and operated to fulfill requests quickly and effectively.
- Prepare to support governance in the host nation following the disintegration or collapse of a regime. Ideally, the civilian departments and agencies of the U.S. government should be prepared to provide advisors and technical personnel at short notice. Alternatively, the U.S. Army's civil affairs units could be expanded and resourced to fulfill this requirement.
- Prepare to fund the establishment of a government within the country, the development of its military, and its reconstruction. Funding mechanisms should assure that funds may be moved flexibly across accounts, expended quickly in response to local contingencies, and monitored effectively by a robust, deployable accounting system.
- Make COIN a primary mission for U.S. military forces, on the same level as large-scale force-on-force combat operations. Military forces should train and exercise to be able to interact with civilian populations and insurgents in complex and ambiguous situations. Joint and service doctrine should treat COIN as a distinct type of political-military operation requiring far closer integration with civilian efforts than would be necessary for large-scale force-on-force combat operations.
- Revise personnel policies to assure retention of skilled personnel in the host country in positions that demand close personal interaction with the indigenous population. Develop legislation to enhance the quality and length of service of U.S. civilian personnel in the country—in effect, a civilian counterpart to the Goldwater-Nichols reform.
- Prepare U.S. conventional military units to partner with corresponding units of indigenous forces. Partnership should imply

continuous association on and off the battlefield, not simply combined operations. It should imply that U.S. military units adapt flexibly to conditions and mentor their counterparts in ways appropriate to their culture and their skill levels.
- Ensure that senior military commanders continuously reexamine the allocation of existing resources (both men and materiel) and that procurement priorities are in line with changing threats on the battlefield.
- Prepare to conduct police work abroad and build foreign police forces on a large scale. The Department of Defense (DoD) or another agency in close coordination with DoD should prepare to introduce large police forces rapidly into areas where governmental authority has deteriorated or collapsed. These police forces should be trained to partner with local police forces at every level, from street patrols to administration at the ministerial level. In Iraq, traditional U.S. military police units were deployed to aid in the COIN effort, but they were trained only in basic skills such as patrolling.
- Refine the ability of brigade-sized formations to conduct joint and combined COIN operations autonomously. These formations should have all the required capabilities, including human-intelligence teams, surveillance systems, translators, and engineer assets. They should be able to obtain intelligence support directly from national assets.
- Develop survivable daylight air platforms with gunship-like characteristics, i.e., comparable to those of the current AC-130 aircraft, to support COIN operations. These characteristics should include long endurance, fine-grained sensing under all light conditions, precise engagement with ordnance suitable for point targets, and robust communications with terminal-attack controllers. These daylight platforms should be survivable against low- and mid-level air-defense weapons.
- Develop the ability to collect intelligence against insurgencies and share it with coalition partners and indigenous forces. Devote special attention to collection of human intelligence, including linguistic skills, interrogation techniques, and devel-

opment of informant networks. Establish procedures and means to share intelligence rapidly with non-U.S. recipients at various levels of initial classification, without compromise of sources and methods.

Acknowledgments

The authors gratefully acknowledge the assistance of General Bryan D. Brown, former Commander, U.S. Special Operations Command (USSOCOM); Ben Riley, Director, Rapid Reaction Technology Office, Office of the Secretary of Defense, Acquisition, Technology, and Logistics; Edward McCallum, Director, Combating Terrorism Technology Support Office, and Richard Higgins, Chief, Irregular Warfare Support Branch, Office of the Secretary of Defense, Special Operations and Low Intensity Conflict; Lt. Gen. (USA) Peter W. Chiarelli, former Commanding General, Multinational Corps-Iraq; Bradford Higgins, former Chief, Joint Planning and Strategy Assessment Cell, U.S. Embassy Iraq; 5th Special Forces Group; Scott Feldmayer, Lincoln Group; Col. (USA, ret.) Joseph Celeski; Col. (USA, ret.) Derek Harvey, Defense Intelligence Agency; Capt. (USN, ret.) Russell McIntyre; Phebe Marr, U.S. Institute for Peace; Kadhim Waeli, Headquarters, U.S. Army Intelligence and Security Command; Col. (USA) Gregory Tubbs, Director Rapid Equipping Force; Lt. Col. (USA, ret.) Lee Gazzano, Rapid Equipping Force; Lt Col (USA) Steve Mannell, U.S. Army Special Operations Command; Lt Col John Cooksey, USSOCOM; Chief Warrant Officer (USAR) Sharon Curcio; Major Ron Watwood and the Psychological Operations Task Force–Baghdad; Headquarters, Army Intelligence Command Directed Studies Office; Col. (USMC) John Stone, G-3, II Marine Expeditionary Force; Col. (USMC)

G. I. Wilson and Col. (USMC) Ed McDaniel, II Marine Expeditionary Force–Fallujah; Michael Newman, Lawrence Livermore National Laboratories; Larry Diamond, Spogli Institute, Stanford University; Provost Marshal's Office, Balad Air Base; Lt. Col. (USA) Pat Crow and Col. (USMC) Jeff Haynes, Future Operations Planning Branch, Multi-National Corps–Iraq; RAND Initiative for Middle East Youth; and RAND colleagues Cheryl Bernard, James Dobbins, John Gordon IV, Adam Grissom, Gene Gritton, Katharine Hall, Stephen Hosmer, Terry Kelley, Walter Perry, Andrew Rathmell, Brian Shannon, Thomas Sullivan, and Fred Wehrey. Of particular importance were the contributions and thoughts of Ambassador Robert Hunter. We offer special thanks to our RAND colleague Richard Mesic and to Ahmed S. Hashim, Naval War College, for reviewing an earlier version of this monograph, as well as to RAND colleague Greg Treverton and David Davis, George Mason University, for completing subsequent reviews. The authors are solely responsible for any errors of fact or opinion.

Abbreviations

AIF	anti-Iraqi Forces
ACR	Armored Cavalry Regiment
CFLCC	Combined Forces Land Component Commander
CIA	Central Intelligence Agency
CJTF-7	Combined Joint Task Force-7
COIN	counterinsurgency
CPA	Coalition Provisional Authority
CPATT	Civilian Police Advisory Training Team
DCI	Director of Central Intelligence
DoD	U.S. Department of Defense
GAO	Government Accountability Office
GPS	Global Positioning System
HAMAS	Harakat al-Muqawama al-Islamiya [Islamic Resistance Movement]
Humvee	High-mobility, multipurpose wheeled vehicle
ICDC	Iraqi Civil Defense Corps
IDP	internally displaced person
IED	improvised explosive device
IMF	International Monetary Fund
IRMO	Iraqi Reconstruction Management Office

IRRF	Iraqi Relief and Reconstruction Funds
KDP	Kurdistan Democratic Party
MEF	Marine Expeditionary Force
MLRS	multiple-launch rocket system
MIPT	Memorial Institute for the Prevention of Terrorism
MNF-I	Multi-National Force–Iraq
MNSTC-I	Multi-National Security Transition Command–Iraq
MOD	Ministry of Defense
MOI	Ministry of the Interior
NATO	North Atlantic Treaty Organization
NCO	non-commissioned officer
NGO	non-governmental organization
ORHA	Office of Reconstruction and Humanitarian Assistance
PRT	Provincial Reconstruction Team
PUK	Patriotic Union of Kurdistan
QRF	Quick Reaction Force
RCIED	remote-controlled improvised explosive device
ROVER	remote operations video-enhanced receiver
SCIRI	Supreme Council for Islamic Revolution in Iraq
SFOR	Stability Force
SIGIR	Special Inspector General for Iraq Reconstruction
UAV	unmanned aerial vehicle
UN	United Nations
UNAMI	United Nations Assistance Mission for Iraq
UNHCR	United Nations High Commissioner for Refugees
UNSC	United Nations Security Council

USA	U.S. Army
USAID	U.S. Agency for International Development
USAR	U.S. Army Reserve
USCENTCOM	U.S. Central Command
USMC	U.S. Marine Corps
USSOCOM	U.S. Special Operations Command
VBIED	vehicle-borne improvised explosive device
WMD	weapons of mass destruction

CHAPTER ONE
Overview of the Conflict in Iraq

The United States is in the fifth year of trying to combat an insurgency that began when it invaded and occupied Iraq. Insurgency against the United States as an occupying power is highly unusual; the only other instance was the Philippine Insurrection of 1898–1903. In the Philippines, however, an insurgency was already under way against the Spanish. Following the defeat of the Spanish, when the United States did not immediately grant the Philippines independence, most of the Filipino insurgents turned their wrath against it. Iraq is the only example of an indigenous insurgency arising in response to a U.S. occupation.

Knowing little about the Philippine Insurrection, most Americans tend to adduce the Vietnam War as an analogy to the Iraq war. Although the two conflicts share some common features, including nationalist sentiment against foreign presence, insurgency against an indigenous government, loss of U.S. reputation abroad, and growing opposition within the United States, Vietnam is a very poor analogy for Iraq. Vietnam had internal divisions, but they were not as fierce as those among Sunni Arabs, Shi'ite Arabs, and Kurds in Iraq. Insurgents in South Vietnam were supported by North Vietnam and were eventually supplanted by regular North Vietnamese forces. These forces fought as entire battalions, accepted sustained combat, and inflicted heavy casualties on U.S. forces. Much of the combat was conducted in extremely difficult terrain—swamps, jungle, and mountains—that impeded the mobility of U.S. forces and compelled them to employ helicopters, which were highly vulnerable to ground fire. The United States had no precision-guided, air-delivered weapons until very late in the war. Ultimately, the war ended with an invasion by North Vietnam, employing conventional forces armed with tanks, artillery, and combat aircraft. Despite the fundamental differences between the Vietnam War and the war in Iraq, U.S. forces might at least have profited from experience in counterinsurgency (COIN) gained from fighting the Viet Cong, but this experience was largely forgotten, except by the Army Special Forces.

Insurgency is a fundamental problem, but the conflict in Iraq is more complicated than simply a revolt against the Iraqi government. The combined facts that the U.S. military has followed a flawed COIN approach and the Iraqi government has been ineffective in exerting its writ of control over the country constituted the early engines of failure. The conflict more nearly resembles a many-sided struggle for power amid the ruins of the Ba'athist state. Broadly speaking, three major groups are involved: Sunni Arabs, who have long dominated Iraq and will not accept an inferior position; Shi'ite Arabs, who are trying to assert a new primacy; and Kurds, who are primarily loyal to the vision—and, increasingly, the reality—of a separate Kurdistan. However, none of these groups is monolithic. Sunni Arabs, on the surface at least, resemble various armed factions along tribal, family, and neighborhood lines in loose collaboration, and Shi'ite Arabs are split into several competing factions with different agendas. Kurds are the most unified group, but even they are split into two parties that fought each other in the past. Most bombing attacks are committed by Sunni extremist groups against U.S. forces and Shi'ites, but Shi'ite groups have also terrorized Sunni civilians and driven them from their homes with assassinations and intimidation. In the absence of strong police forces, criminal activity, especially robbery, extortion, and kidnapping, have become a virulent strain. Hostility to foreign influence is one of the few common threads among Iraqi Arabs.

The Ba'athist Regime

After a succession of military coups, Saddam Hussein seized power in 1979 and imposed a dictatorship dominated by Sunni Arabs. He ultimately developed a dictatorship exercised through the Ba'ath Party and characterized by sycophantic idolization of himself, government intrusion into every element of society, a virulently anti-Semitic policy, glorification of military power, and restless aggression. The Ba'athist regime perpetuated Sunni dominance and ultimately degenerated into a gangster state dominated not only by Sunni Arabs, but also more narrowly by members of Saddam's extended family and followers from his hometown of Tikrit, north of Baghdad. The regime committed mass atrocities against separatist Kurds and Shi'ite Arabs suspected of being in league with Shi'ite Iran.

Saddam Hussein plunged Iraq into disastrous wars and repressed revolts with extreme brutality, leaving a legacy of fear and resentment among Kurds and Shi'ite Arabs. His refusal to renounce weapons of mass destruction led the United Nations (UN) Security Council to impose sanctions that impoverished ordinary Iraqis outside the ruling elite. His highly corrupt administration allowed the country's infrastructure to deteriorate. At the same time, it distorted the economy through massive government subsidies of necessities, including food and fuel. As a result, Iraq emerged from the Ba'athist dictatorship divided by ethnic and sectarian tensions, burdened with inefficient government subsidies, suffering from decrepit infrastructure, and accustomed to understanding politics as a brutal winner-take-all contest.

In 1980, Saddam's forces attacked the newly founded Islamic Republic of Iran. He apparently expected to acquire the Arab-inhabited, oil-rich province of Khuzestan. At the same time, he feared the threat posed by Ayatollah Khomeini's revolutionary Islam, especially among Iraq's large Shi'ite population. In an unusual convergence of policy, the United States and the Soviet Union supported Iraq despite its aggression, because revolutionary Iran appeared to be more threatening than the Ba'athist regime. The Iran-Iraq War became a bloody stalemate reminiscent of the Western Front during World War I. After concluding a ceasefire with Iran in 1988, Saddam launched a large-scale offensive against Kurdish forces, which had allied with Iran. The Iran-Iraq War left both countries exhausted and nearly bankrupt despite their oil revenues.

In mid-1990, Saddam accused Kuwait of slant-drilling into Iraqi fields and demanded relief from war debts. When Kuwait refused his demand, Saddam invaded the country and proclaimed it to be the nineteenth province of Iraq. In response, the United States and Saudi Arabia assembled a large coalition of Western and regional powers pursuant to a resolution of the UN Security Council. Coalition forces, primarily U.S. forces, after five weeks of air attacks, freed Kuwait of Iraqi forces, which seldom offered more than token resistance.

In the wake of Saddam Hussein's defeat in Kuwait, Kurds in the north and Shi'ite Arabs in the south rose up against his regime. Pursuant to another Security Council resolution, the United States formed a much smaller Western coalition to aid Kurds who had fled across the border to Turkey. Under pressure from this coalition, Saddam Hussein withdrew troops from Kurdish-inhabited areas other than Kirkuk, allowing the Kurds to enjoy autonomy. Unfortunately for their cause, the Kurds fell into a fratricidal conflict between the Patriotic Union of Kurdistan (PUK) led by Jalal Talabani and the Kurdistan Democratic Party (KDP) led by Massoud Barzani, the latter supported by Iraqi forces. In September 1998, the United States brokered a ceasefire between these Kurdish factions.

From Saddam Hussein's point of view, the Shi'ite revolt in southern Iraq was the most dangerous threat to his regime, and even to the continued existence of Iraq, because Shi'ite Arabs far outnumber Sunni Arabs. The Supreme Council for Islamic Revolution in Iraq (SCIRI), an organization based in Tehran, supported Iraqi Shi'ites financially and through the Badr Corps, a Shi'ite militia that had fought alongside Iranian forces during the war. To suppress revolt among marsh Arabs, Saddam Hussein drained the marshes between the Tigris and Euphrates Rivers, causing an ecological disaster. He suppressed the Shi'ite revolt with great brutality, but even so, it continued to smolder. After encouraging this revolt, the United States took no action to prevent its suppression, leaving a legacy of mistrust among the Shi'ite Arabs.

The Invasion of Iraq

In 2002, the Bush Administration announced its determination to invade Iraq if necessary to prevent Saddam Hussein from gaining weapons of mass destruction, which he could use for conquest or could provide to terrorists. In addition, it expected that Iraq would become a peaceful, democratic state that would serve as an example for the region. In his State of the Union Address in January 2003, President Bush said:

> Year after year, Saddam Hussein has gone to elaborate lengths, spent enormous sums, taken great risks to build and keep weapons of mass destruction. But why? The only possible explanation, the only possible use he could have for these weapons, is to dominate, intimidate, or attack.
>
> With nuclear arms or a full arsenal of chemical and biological weapons, Saddam Hussein could resume his ambitions of conquest in the Middle East and create deadly havoc in that region. And this Congress and the American people must recognize another threat. Evidence from intelligence sources, secret communications, and statements by people now in custody reveal that Saddam Hussein aids and protects terrorists, including members of al Qaeda. Secretly and without fingerprints, he could provide one of his hidden weapons to terrorists, or help them develop their own.[1]

In a speech at the UN in February 2003, U.S. Secretary of State Colin Powell stated that Iraq sought weapons of mass destruction and was involved in terrorism. He described a "sinister nexus between Iraq and the al Qaeda terrorist network" and said that "ambition and hatred are enough to bring Iraq and al Qaeda together, enough so that al Qaeda could learn how to build more-sophisticated bombs and learn

---

[1] President George W. Bush, State of the Union Address (Washington, D.C., January 28, 2003). As of October 1, 2007: http:/www.whitehouse.gov/news/releases/2003/01/print/20030128-19.html.

how to forge documents, and enough so that al Qaeda could turn to Iraq for help in acquiring expertise on weapons of mass destruction."[2]

On March 17, 2003, President Bush delivered an ultimatum to Iraq, stating that the United States would initiate military conflict unless Saddam Hussein and his sons left Iraq within 48 hours. In this ultimatum, President Bush said:

> The regime has a history of reckless aggression in the Middle East. It has a deep hatred of America and our friends. And it has aided, trained, and harbored terrorists, including operatives of al Qaeda.
>
> The danger is clear: using chemical, biological, or, one day, nuclear weapons, obtained with the help of Iraq, the terrorists could fulfill their stated ambitions and kill thousands of innocent people in our country, or any other.[3]

Considering Iraq's defeat in the previous Persian Gulf War, Saddam Hussein appeared surprisingly insouciant in the face of U.S. demands. According to research conducted after the 2003 invasion, he had convinced himself that the United States lacked resolve and was afraid to risk an invasion.[4] Events after the previous war, such as the U.S. defeat in Somalia, confirmed his impression that the United States was irresolute and highly sensitive to casualties. Even after abandoning his programs to develop weapons of mass destruction, Saddam still refused to cooperate fully with UN inspectors, for reasons of pres-

---

[2] Colin Powell, "Address to the U.N. Security Council" (New York: United Nations, February 5, 2003). As of October 1, 2007: http://www.whitehouse.gov/news/releases/2003/02/print/20030205-1.html.

[3] "President Says Saddam Must Leave Iraq Within 48 Hours" (Washington, D.C.: The Cross Hall, White House, March 17, 2003, p. 1.) As of October 1, 2007: http://www.whitehouse.gov/news/releases/2003/03/print/20030317-7.html.

[4] See "Saddam's Distorted Worldview," in Kevin M. Woods et al., *Iraqi Perspectives Project: A View of Iraqi Freedom from Saddam's Senior Leadership* (Norfolk, Va.: Joint Center for Operational Analysis and Lessons Learned, U.S. Joint Forces Command, 2006).

tige. He failed to understand that the United States interpreted his lack of cooperation as proof that the programs still existed.

For several reasons, U.S. forces encountered little resistance from the Iraqi Army during the invasion: Saddam regarded Kurds and Shi'ites backed by Iran as more immediate threats than an unlikely U.S. invasion. Therefore, he kept most of his forces opposite Kurds and Iranians, leaving the invasion corridor through Kuwait to the vicinity of Baghdad largely unprotected. Baghdad was defended by Republican Guard divisions deployed around the city, but even they offered only sporadic resistance. Saddam and his two sons issued amateurish and confusing orders to their military commanders, who were not allowed to exercise any initiative. The Iraqi Army was neglected, demoralized, and poorly trained even by regional standards. Moreover, the Iraqi soldiers knew from experience that U.S. forces were overwhelmingly superior, and therefore most of them deserted before making contact.

Prior to the invasion, Saddam Hussein developed several paramilitary forces that later fed into the insurgency. He developed a large militia called the Al-Quds (Jerusalem) Army, ostensibly to help defeat Israel, but actually to defend areas within Iraq where he feared unrest. Although its troops numbered in the hundreds of thousands, the Al-Quds Army had negligible military value. It was commanded by Ba'athist politicians who were almost untrained and were by no means equipped to confront any serious military force. Saddam also developed a smaller, but more lethal force called Fedayeen Saddam. He initially created this force to repress Shi'ite Arabs and Kurds but subsequently gave it a security mission against all enemies of the regime. Fedayeen Saddam operated training camps that hosted volunteers from regional Arab countries, some of whom fought against U.S. troops during the invasion. Prior to the invasion, Saddam also gave orders to store food, fuel, and ammunition in safe places, including mosques and schools.

In contrast to Iraqi regular forces, substantial numbers of irregular forces, especially Fedayeen Saddam, offered resistance to the invasion. The Fedayeen were lightly equipped and poorly trained, yet they threw themselves against U.S. units. They fought with complete disregard for the laws of warfare; for example, they used mosques and hospitals for military purposes and concealed themselves among civilians, offering a preview of the insurgency that would follow:

> While the Fedayeen were not part of a deliberate plan to carry on a guerrilla war in the event the regime was toppled, it provided much of the wherewithal for an insurgency: thousands of committed fighters, decentralized command and control systems, and massive caches of arms.[5]

The Ba'ath Party was secularist in its inception, but Saddam Hussein subsequently portrayed himself as an obedient follower of the prophet and appealed to Islamic sentiment whenever it suited his purpose. He appealed to Islam during his protracted conflict with the United States and postured as the champion of Islam against Jewish encroachment in Palestine. As a result, the Ba'ath Party ceased being secular and adopted at least a façade of Islamic faith. After the invasion, former Ba'athist leaders appealed to both national and Islamic sentiment in opposition to an infidel invader. Sunni mosques became centers of opposition, where religious leaders inspired the faithful to oppose the occupation.

During the last decade of Ba'athist rule, the Kurds enjoyed autonomy from Baghdad and secured their territory with their own militia, the Pesh Merga, which in effect became a Kurdish Army. During the invasion of Iraq, the U.S. deployed Army Special Forces and small conventional forces into Kurdish-controlled areas. Special Forces worked closely with the Pesh Merga to attack Ansar al-Islam and the Iraqi Army. Supported by U.S. air forces, the Pesh Merga broke through Iraqi Army defenses and occupied the cities of Mosul and Kirkuk.

Knowing that it would soon occupy Iraq, the United States avoided bombing infrastructure. Indeed, U.S. planners were more concerned with protecting Iraqi infrastructure than with destroying it. They feared that Saddam Hussein would order petroleum facilities to be ignited, as he had done in Kuwait during the Persian Gulf War, or

---

[5] Michael R. Gordon and Lt. Gen. (USMC, ret.) Bernard E. Trainor, *Cobra II, The Inside Story of the Invasion and Occupation of Iraq* (New York: Pantheon Books, 2006, p. 505).

would destroy a major dam on the Euphrates River. However, Saddam had no intention of destroying these facilities, because he did not anticipate losing them. He did not even order bridges to be destroyed, presuming that his own forces would need them.

The Occupation of Iraq

In defeat, the Ba'athist regime collapsed quickly and completely. Suddenly no longer a centrally controlled, one-party dictatorship, Iraq became ungoverned space, lacking basic services and security for its citizens. Iraqi civilians looted government offices and installations of everything movable, including electrical wiring and plumbing fixtures, leaving only shells behind. Two months after the invasion, Ambassador L. Paul ("Jerry") Bremer arrived in Baghdad. He subsequently recalled that he had been extensively briefed on Iraq. "But," he said, "nobody had given me a sense of how utterly *broken* this country was" [emphasis in the original].[6]

Through the summer of 2003, resistance to the occupation stayed at a low, relatively tolerable level. There were two main sources of this resistance: remnants of the Ba'athist regime and extremists, especially foreign fighters from other countries in the Middle East. Because Iraqi armed forces had deserted rather than being defeated in battle and surrendering, Iraq was thickly sown with former soldiers, weapons, and munitions. Although Saddam and his sons were still at large, they apparently did not exercise much control over the insurgency. Former Ba'athists, especially Army officers and members of the security apparatus, organized resistance at local levels. This resistance became a serious problem by the fall of 2003. In addition, insurgents were supplemented by foreign fighters from other Middle Eastern countries. During the last months of his regime, Saddam had welcomed foreign volunteers into Iraq, and some, especially Syrians, fought against U.S. forces during the invasion. After the fall of Baghdad, more foreigners

---

[6] L. Paul Bremer III, with Malcolm McConnell, *My Year in Iraq, The Struggle to Build a Future of Hope* (New York: Simon & Schuster, 2006, p. 18).

came to resist the U.S. occupation. However, their numbers remained small compared with the number of Iraqi insurgents, who were almost exclusively Sunni Arabs.

On July 22, 2003, acting on a tip, U.S. forces surrounded a house in Mosul where Saddam's sons Uday and Qusay were hiding. The sons offered resistance and died, together with their bodyguards, in a gun battle. On December 13, U.S. forces captured Saddam Hussein, who was found in a cramped underground hiding place near Tikrit. The former dictator was disheveled and appeared to be disoriented. For several weeks after his capture, there was a lull in insurgent attacks, but they then resumed their former tempo.

Considering the circumstances of his capture, Saddam was probably not leading the insurgency, although he helped promote it. On November 5, 2006, after a year-long trial, the Iraqi Special Tribunal sentenced him to death by hanging for having ordered the killing of civilians in the village of Dujail following a 1982 attempt on his life.

The First Priority: Setting Up a Constitutional Government

When Ambassador Bremer met with leading Iraqi politicians on May 16, 2003, they advised him that a new government was urgently needed. Jalal Talabani said, "While we sincerely thank the coalition for all its efforts, we have to warn against squandering a military victory by not conducting a rapid coordinated effort to form a new government."[7]

On July 13, 2003, Bremer announced formation of the Iraqi Governing Council, composed of 25 members chosen by the coalition. The council chose as its first president Ibrahim al-Jaafari, a leader in the Shi'ite Da'wa Party, whose members included Abd al-Aziz al-Hakim, the leader of SCIRI; Masud Barzani, head of the Kurdistan Democratic Party; and Jalal Talabani, head of the Patriotic Union of Kurdistan. The new council thus included prominent leaders of Shi'ite Arabs and Kurds, but not of Sunni Arabs, who opposed it.

---

[7] Bremer, 2006, p. 48.

The council approved an interim constitution, known as the Law of Administration for the State of Iraq for the Transitional Period, drafted by the Coalition Provisional Authority (CPA) in coordination with the council. This interim constitution provided that government "shall be republican, federal, democratic, and pluralistic" (Article 4); that "Islam is the official religion of the State and is to be considered a source of legislation" (Article 7); that the "Arabic language and the Kurdish language are the two official languages" (Article 9); and that natural resources shall be managed "distributing the revenues resulting from their sale through the national budget in an equitable manner proportional to the distribution of population throughout the country" (Article 25(E)). A "fully sovereign Iraqi Interim Government" was to take power on June 30, 2004 (Article 2(B)), but Bremer relinquished his authority two days earlier to assure his safe departure from the country.[8]

The Spiral Downward Begins (Spring 2004)

In the spring of 2004, the United States conducted large military operations against Sunni Arab insurgents in Fallujah and against a newly formed Shi'ite militia led by the stridently nationalistic Muqtada al-Sadr. Sunni Arab inhabitants of Fallujah were especially hostile to occupation, and U.S. forces had little presence in this city. On March 31, insurgents ambushed a small civilian convoy protected by four Blackwater[9] security personnel on a street in Fallujah. The incident revealed a high level of ferocity and hatred toward Americans. Crowds savagely abused the four bodies, burning them beyond recognition and parading the charred remains through the streets. Two of the remains were hung from the girders of a bridge while citizens of Fallujah celebrated below. Militants displayed a sign reading "Fallujah is the graveyard of Americans" and brandished weapons in a show of defiance. In

---

[8] Ibid., pp. 392–395.

[9] Blackwater is a private military contractor currently providing security services to U.S. government agencies in Iraq.

response, the United States ordered Marine forces to occupy Fallujah but stopped the offensive when members of the Iraqi Governing Council threatened to resign in protest. Finally, on November 7, 2004, U.S. Army and Marine Corps forces initiated an operation to seize Fallujah. Insurgents had spent the summer preparing for an assault that both sides knew was coming. As a result, U.S. forces had to destroy numerous strongpoints, inflicting considerable damage on the city. There were few civilian casualties, because most of the inhabitants had been warned of the impending assault and fled the city before it began, but reconstruction was a slow process.

Bremer had long viewed al-Sadr with alarm and urged action against his organization. In March 2004, the CPA banned al-Sadr's newspaper, his primary means of communicating with his supporters. He responded by seizing control of Najaf, a city south of Baghdad famous for containing the shrine of Imam Ali, a revered figure in Shi'ite history. U.S. forces clashed with al-Sadr's militia in widely scattered locations, including Basra, Baghdad, and Nasiriyah, but the most important battle was in Najaf. Many pious Shi'ites bring the bodies of deceased relatives to this holy city for burial, and as a result, a vast cemetery extends north of the shrine. U.S. forces fought through this cemetery with heavy armor until they approached and cautiously secured the venerated shrine. Under pressure from the council, the United States accepted mediation by the most influential Shi'ite figure, the Ayatollah Ali al-Hussein al-Sistani, who brokered a ceasefire that denied al-Sadr control over Najaf but left his power otherwise unimpaired. On June 11, al-Sadr publicly urged his followers to adhere to the ceasefire, at the same time endorsing an Iraqi interim government that would supplant the U.S. occupation. With this decision, he began a political career to complement his role as a militia leader.

In the absence of a clear overall COIN strategy, coalition forces focused on tactical matters, executing door-to-door raids mixed with presence patrols in Baghdad and other cities; both approaches proved increasingly intermittent and ineffective over time. The coalition also tried to focus on securing its key main supply routes. These routes had become increasingly predictable to the insurgents. By the time senior U.S. military authorities finally acknowledged that their forces were engaged in a counterinsurgency in April 2004, the coalition had already become overly concentrated on the tactical and technical challenge posed by roadside improvised explosive devices (IEDs).

Of the four key Multi-National Force–Iraq (MNF-I) objectives in the summer of 2004, two—supporting the Iraqi government and securing lines of operation—took priority over the provision of essential services to the population and strategic communications. Unfortunately, the security of lines of operation was mostly for the benefit of the coalition forces and senior Iraqi officials, and strategic communications were often directed at supporting governance and Iraqi unity rather than at reducing insurgent recruiting and informing the Iraqi citizenry of the goals of coalition military operations. Some leaders, including the 1st Cavalry Division Commander, experimented with the provision of economic services in Sadr City as a way to quell radical militia operations and recruiting, but the effort was never proven to have a lasting effect. The end result was that militia leaders took credit anyway, and the switch away from kinetic operations only appeared to encourage increased insurgent and militia adventurism in 2005.

Nearly every day, clandestine groups abducted their adversaries, tortured them, murdered them, and threw their bodies onto the street. Terrorism impeded reconstruction in several ways. It drove most of the international agencies and non-governmental organizations (NGOs) out of Iraq, prompted donor countries to reconsider making contributions, discouraged private companies from investing, compelled the United States to divert funds toward security, continued to drive out the remnants of the Iraqi middle class, and disrupted projects that were under way in Iraq.

Benchmark One: Holding Iraqi Elections

Bremer would have preferred to schedule national elections at a later date, but Shi'ite Arabs led by al-Sistani demanded early elections. On January 14, 2004, tens of thousands of Shi'ites demonstrated in support of this demand. As a result, Iraq's interim constitution stipulated that elections to the Transitional National Assembly would take place not later than the end of January 2005. This election was held on January 30, 2005, using a system of proportional representation in which voters chose among lists, and seats were allocated in proportion to their choices.[10] The United Iraqi Alliance, based on the Council for Islamic Revolution in Iraq and the Da'wa Party, appealed to Shi'ite Arabs. The Kurdistan Alliance, combining the KDP and the PUK, appealed to Kurds. The only major list not defined in ethnic or sectarian terms was the Iraqi List, led by the interim prime minister, Iyad al-Allawi, and built on his Iraqi National Accord Party. Sunni voters largely boycotted the elections on the urging of the Iraqi Muslim Scholars Association. The United Iraqi Alliance won 140 seats; the Kurdistan Alliance won 75 seats; the Iraqi List won 40 seats; and Sunni Arabs won only 17 seats spread over several lists. Ibrahim al-Jaafari from the Shi'ite Da'wa Party became Prime Minister.

On October 15, 2005, Iraqis voted to accept Iraq's draft constitution in a national referendum. On December 15, 2005, they voted for the Council of Representatives under this new constitution. This time, Sunni Arabs participated in the election. The Iraqi Electoral Commission announced that 10.9 million voters had cast ballots, representing about 70 percent of registered voters. More than previously, sectarian- and ethnically based parties dominated the results. The United Iraqi Alliance won 128 seats; the Kurdistan Alliance won 53 seats; the Iraqi List fell to 25 seats; and the Sunni-based Iraqi Concord Front won 44 seats. Included in the United Iraqi Alliance was al-Sadr's movement, which obtained 30 seats.

Following the December 2005 election, the United Iraqi Alliance again offered Ibrahim al-Jaafari as Prime Minister, but Kurds, Sunnis, and secular Iraqis all opposed his nomination. With support from al-Sadr, the Da'wa Party leader, Nouri al-Maliki, was named Prime Minister in February 2006. Negotiations among the parties continued for several months until June 9, 2006, when they at last agreed on a full cabinet slate, including the powerful ministers of defense and the inte-

---

[10] For a summary of election results, see Kenneth Katzman, *Iraq: Elections, Government, and Constitution* (Washington, D.C.: Congressional Research Service, 2006).

rior.[11] In the distribution of ministerial posts, 21 went to Shi'ite Arabs, eight to Sunni Arabs, seven to Kurds, and one to a Christian.

The insurgents' success had both political and military effects. The growing insurgency prevented the Iraqi government from exerting its writ of control across Iraq. The relationship between insurgent groups remains to this day a complex milieu of Sunni Arab insurgents, Shi'ia militia, criminal gangs, foreigners, and other opportunists who conduct business at a transactional level—which is why U.S. efforts to split or wedge these groups and their leaders from one another have proven so difficult.

Islamic Extremist and Sectarian Violence Begin

Foreign fighters initially lacked a formal link to al Qaeda, whose leaders were presumably hiding in southern Afghanistan or northwestern Pakistan. But on October 17, 2004, Jordanian terrorist Abu Musab al-Zarqawi (Ahmed Fadil al-Khalayleh) announced his allegiance to Osama bin Laden. His organization, the Unity and Jihad Group, became known as al Qaeda in Mesopotamia (also translated as al Qaeda in the Land of the Two Rivers). In January 2006, al Qaeda was subsumed into a larger grouping called the Majilis Shura al-Mujahideen fi al-Iraq (Mujahideen Shura Council in Iraq). On June 7, 2006, U.S. F-16 aircraft dropped several laser-guided 500-lb bombs on a house near the town of Baquba, killing al-Zarqawi in the blasts. Prime Minister al-Maliki announced that this strike was based on information that residents in the area provided to Iraqi intelligence.[12] On October 15, 2006, the Mujahideen Shura Council announced formation of an Islamic State of Iraq comprising the provinces of al Anbar,

---

[11] Both of these ministers were former Army officers. The new minister of defense was Abdul Qadar Mohammed Jassim, a Sunni Arab and former Army general, who had been jailed for opposing the invasion of Kuwait. The new minister of the interior was Jawad al-Bolani, a Shi'ite Arab and retired Army colonel.

[12] "Abu Musab Al-Zarqawi, Leader of Al Qaeda In Iraq, Has Been Killed," *The New York Times on the Web*, June 8, 2006. As of October 1, 2007: http://topics.nytimes.com/top/reference/timestopics/people/z/abu_musab_al_zarqawi/index.html?8qa.

Diyala, Kirkuk, Ninawa, Salahad Din, and parts of Babel and Wasit. The council claimed to be taking this act in response to Kurds and Shi'ite Arabs securing semi-autonomous regions within Iraq. Islamic extremists continued attacking Shi'ite Arabs, whom they portrayed as apostates in league with foreign occupiers.

By early 2006, sectarian violence was escalating in areas where Sunni and Shi'ite Arabs were mixed, especially the Baghdad area. In some neighborhoods, Iraqis relied on militias and less-formal organizations for security; however, these were increasingly outlawed by U.S. and Iraqi security forces. In several areas, Sunni and Shi'ite Arabs began to relocate along sectarian lines, amid violence reminiscent of the ethnic cleansing that occurred in the Balkans, especially Bosnia. In congressional testimony on August 3, 2006, the U.S. Central Command commander, General (USA) John P. Abizaid, said, "I believe that the sectarian violence is probably as bad as I've seen it in Baghdad in particular, and that, if not stopped, it is possible that Iraq could move towards civil war."[13] Abizaid and the Chairman of the Joint Chiefs of Staff, General (USMC) Peter Pace, both said that they had not anticipated that sectarian violence would rise to such a level.

On February 22, 2006, foreign extremists overcame guards and destroyed the golden dome of the Ali al-Hadi Mosque in Samara, 60 miles north of Baghdad. The tombs of Ali al-Hadi and Hassan al-Askari, two of the original 12 imams, are located in this mosque, an object of veneration for Shi'ites. Its destruction triggered Shi'ite Arab demonstrations and attacks on Sunni mosques in Baghdad and Basra. Ayatollah al-Sistani released a statement saying, "If the government's security forces cannot provide the necessary protection, the believers will do it."[14]

By early 2006, U.S. officials estimated that Shi'ite militias were killing more people than Sunni insurgents were and were becoming

---

[13] Hearing before the Senate Armed Services Committee, Washington, D.C., August 3, 2006. General Abizaid was responding to a question from Senator Carl Levin (Democrat, Michigan) as to whether Iraq might be sliding toward civil war. As of October 1, 2007: http://www.pbs.org/newshour/bb/middle_east/july-dec06/military_08-03.html.

[14] Robert F. Worth, "Blast Destroys Shrine in Iraq, Setting Off Sectarian Fury," *The New York Times*, February 22, 2006.

the greatest challenge to the Iraqi government.[15] The militias were represented politically within the government, and they infiltrated Iraqi police forces.

A U.S. Approach Hesitantly Unfolds

In late 2005, the U.S. National Security Council defined conditions in the short, mid-, and long term that would constitute "victory in Iraq."[16] Over the short term, Iraq would be making "steady progress" in fighting terrorists, neutralizing the insurgency, building democratic institutions, maintaining security, and tackling key economic reforms. In the mid-term, Iraq would be taking the lead in these areas and would be well on its way to achieving its economic potential. In the long term, Iraq would become a peaceful, united, stable, democratic, and secure country that would be a partner in the global war on terror, an engine for regional economic growth, and proof of the fruits of democratic governance. The National Security Council also defined metrics to measure progress in political process, security, and economic growth. For example, security was to be measured by the quantity and quality of Iraqi units, actionable intelligence received from Iraqis, the percentage of operations conducted by Iraqi units without assistance, the number of car bombs intercepted, offensive operations by friendly forces, and the number of contacts initiated by coalition forces rather than by the enemy.

Within the United States, public support for the war in Iraq steadily eroded over time. In late April 2003, more than 80 percent of survey respondents said things were going well, but by March 2006, 60 percent said things were going badly.[17] During the same period,

---

[15] Jonathan Finer, "Threat of Shi'ite Militias Now Seen as Iraq's Most Critical Challenge," *The Washington Post*, April 8, 2006, p. 1.

[16] National Security Council, *National Strategy for Victory in Iraq* (Washington, D.C.: National Security Council, 2005, p. 3).

[17] In a poll conducted by Gallup, Cable News Network, and *USA Today* on March 10–12, 2006, 6 percent of the respondents said things were going very well, 32 percent said they were going moderately well, 32 percent said moderately badly, and 28 percent said very badly.

the U.S. public was about equally divided over whether U.S. troops should stay as long as it took to assure a stable democracy or should leave even if the country was not completely stable. By early 2006, a majority of respondents thought the United States should have stayed out of Iraq.[18]

In September 2006, General Pace formed a study group to consider changes in military strategy in Iraq; the group's report was expected in December.[19] On November 6, 2006, one day prior to the midterm elections for Congress, Secretary of Defense Donald Rumsfeld sent a memorandum to the White House saying, "In my view, it is time for a major adjustment. Clearly, what U.S. forces are currently doing in Iraq is not working well enough or fast enough."[20] Rumsfeld offered a list of illustrative options that included benchmarks to get the Iraqi government moving, a significant increase in assistance to Iraqi forces, help from the Department of Defense (DoD) for key Iraqi ministries, and modest withdrawals of U.S. forces so that Iraqis would take more responsibility.

In the midterm elections, the Democratic Party won control of both Houses of Congress, in part due to the unpopularity of the war. On the following day, President Bush accepted Rumsfeld's resignation and announced the nomination of Robert Gates as his successor. On December 6, the Iraq Study Group issued a report on the war that concluded that the situation in Iraq was deteriorating, time was running out, and current U.S. policy was not working. The Study Group

---

[18] The Columbia Broadcasting System, sometimes in conjunction with *The New York Times*, repeatedly asked this question: "Looking back, do you think the United States did the right thing in getting involved in a military conflict with Iraq or should the United States have stayed out?" On March 26–27, 2003, 69 percent of the respondents said "right thing," while 25 percent said "should have stayed out." By early 2005, respondents were about evenly divided, and by fall 2005, most respondents said "should have stayed out." In a poll conducted on April 6–9, 2006, three years after the invasion, 43 percent of the respondents said "right thing," and 53 percent said "should have stayed out."

[19] Elaine M. Grossman, "Pace Group to Put Forth Iraq Strategy Alternatives by Mid-December" (*Inside the Pentagon*, November 9, 2006).

[20] Michael R. Gordon and David S. Cloud, "Rumsfeld's Memo on Iraq Proposed Major Change" (*The New York Times*, December 3, 2006). For the text of the memorandum, see "Rumsfeld's Memo of Options for Iraq War" (*The New York Times*, December 3, 2006).

offered 79 recommendations to help attain the goal of an Iraq that could govern, sustain, and defend itself. It recommended that if the Iraqi government did not make substantial progress, the United States should reduce its support.[21] On November 14, President Bush had directed an internal review of Iraq policy, under the oversight of National Security Advisor Stephen J. Hadley, to be completed in mid-December.[22] President Bush would thus have at least three sources of advice to help develop a new strategy: Pace's study, the Iraq Study Group Report, and Hadley's internal review.

To the surprise of many, President Bush rejected the Iraq Study Group's recommendations and decided to introduce additional U.S. forces into Iraq, a so-called "surge." The purpose of this policy change, which would bring the number of U.S. troops in the country to more than 150,000 by late spring 2007, was to secure the Baghdad area. In theory, decreasing the level of violence in the capital would facilitate a political solution between Iraq's various groups, a solution that had eluded the United States since its initial entry into the country in 2003.

In early 2007, Army General David Petraeus became the commander of U.S. forces in Iraq. Bringing with him a number of Army colonels who had COIN experience to form a key inner circle of staff in Baghdad, Petraeus was charged with implementing the "surge." For the first time since the invasion in 2003, the mission of U.S. forces, at least in the Baghdad area, would be to provide security for the population. The hope was that the increased numbers of U.S. and Iraqi troops there would break the control various militia groups had in many neighborhoods of the city. At the time of this writing (summer 2007), the effects of this increase in troop strength—and the mission of protecting the population—are still uncertain.

---

[21] "RECOMMENDATION 21: If the Iraqi government does not make substantial progress toward the achievement of milestones on national reconciliation, security, and governance, the United States should reduce its political, military, or economic support for the Iraqi government," James A. Baker III and Lee H. Hamilton, Co-Chairs, *The Iraq Study Group Report* (New York: Vintage Books, 2006, p. 61).

[22] Robin Wright, "Bush Initiates Iraq Policy Review Separate from Baker Group's" (*The Washington Post*, November 15, 2006).

CHAPTER TWO
Armed Groups in Iraq

Although primarily characterized as an insurgency, the conflict in Iraq involves a mixture of armed groups with conflicting goals. Were insurgency the only challenge, U.S. and Iraqi government forces might at least contain it, but multiple challenges of separatists, insurgents, extremists, militias, and criminals threaten to destroy the country.

At least through 2005, most of the violence in Iraq was caused by a Sunni-dominated insurgency against U.S. forces who were seen as occupiers even after the notional return of sovereignty in June 2004. The insurgency originated among Ba'athist remnants, especially the ruling family and its enforcers, who could expect to find no place in the new Iraq. However, it gained wide support among Sunni Arabs and continued to grow even when many from the former Ba'athist leadership were killed or captured in 2003–2004. Once stridently secular, the Ba'athist regime had later evoked Islam to increase its popularity. At least some of the insurgents were also strongly Islamic, especially in outlying towns of the so-called Sunni Triangle north and west of Baghdad. The Sunni insurgents' combination of nationalist sentiment and Islamic fervor was shared by the Shi'ite Arab movement led by Muqtada al-Sadr. Despite these similar motivations, Sunni insurgents and al-Sadr's movement were divided by their attitudes toward Iran and the fact that they represented different branches of Islam that have quarreled for centuries. Like most Shi'ite Arab leaders, al-Sadr was friendly toward the clerical regime in Iran, which Sunni Arabs deeply distrusted.

Overview

Violence in Iraq currently involves separatists, insurgents, violent extremists, Arab Shi'ite militias, and criminals.

- **Separatists and sectarianism.** Separatism and sectarianism compound Iraq's problems and appear to be increasing. Kurds do not regard themselves as Iraqis; they stay within Iraq as a matter of convenience. Most Arabs do consider themselves Iraqis and would prefer to maintain a single state, but their leadership shows little commitment to a unified pluralistic government. The Supreme Council for Islamic Revolution in Iraq (SCIRI) publicly advocates autonomous regions and envisions a large Shi'ite-dominated region in southern Iraq. In contrast, Sunni leadership has little interest in creating an autonomous Sunni region, if only because the region would not contain producing oil fields. However, the extremist Mujahideen Shura Council in Iraq recently announced the creation of a new Islamic state encompassing Sunni-inhabited areas.
- **Insurgents.** The conflict centers on an insurgency along a sectarian and ethnic divide, i.e., Sunni Arab opposition to an Iraqi government dominated by Shi'ite Arabs and Kurds, who are supported by U.S. forces. Countering this insurgency is fundamental, because success would allow the Iraqi government to concentrate on other urgent problems. To succeed, the Iraqi government must be perceived as impartial and able to protect all of its citizens. Creating such a perception is extremely difficult amid escalating sectarian violence, especially when government ministries are involved with sectarian militias.
- **Violent extremists.** Extremists gravitate to the conflict for various reasons. On an overall level, it fits into a vision of protecting Muslim countries against foreign domination. On a personal level, it offers an outlet for resentment and a chance to attain personal redemption. In addition, many terrorist leaders are Salafist (fundamentalist) Sunnis who deliberately incite sectarian violence by attacking Shi'ite civilians.

- **Shi'ite Arab militias.** Many Arab Shi'ites depend on militias more than on Iraqi government forces for security. The militia leaders exert strong influence within the government, which refuses to curb their activities. The Badr Organization was created during the Iran-Iraq War, while the much larger Mahdi Army emerged during the U.S. occupation. The Mahdi Army combines security functions with social services, much like Hezbullah in Lebanon, thus becoming a quasi-state within a state. It appears to be linked with Shi'ite death squads that abduct, torture, and kill Sunni Arabs in the Baghdad area.
- **Criminals.** Criminality continues to plague the country, and criminals hire out their services to enemies of the Iraqi government. Although few crime statistics are kept, it appears that many Iraqis consider criminality to be the greatest threat in their daily lives. The government's inability to combat crime diminishes its reputation and its appeal to loyalty.

At the time of the invasion, Kurds enjoyed autonomy, protected by their own militia, the Pesh Merga. They gave their allegiance to Kurdistan, not to the Iraqi government. Sunni Arabs resisted U.S. occupation from the outset, using unconventional forces, initially through the Fedayeen Saddam and subsequently through amorphous insurgent organizations. Following the invasion, violent extremists, many from neighboring countries, committed terrorist acts not only against U.S. and Iraqi government forces, but also against Shi'ite Arab civilians, in an attempt to incite sectarian violence. Partially in response to these attacks, Shi'ite Arabs turned increasingly to militias for protection. The most powerful Shi'ite militia was the Mahdi Army led by Muqtada al-Sadr. His followers stood for election and accepted ministerial posts in the Iraqi government but remained bitterly opposed to the U.S. presence.

After participating in the national elections of December 2005, Sunni Arabs formed a minority in an Iraqi government dominated by Kurds and Shi'ite Arabs. Secular and nonaligned parties dwindled into insignificance, and Iraqi politics centered on negotiations among blocs defined by religion (Sunni or Shi'ite) and ethnic origin (Kurd). Shi'ite militias infiltrated government institutions, especially the Ministry of the Interior, and used these institutions to attack Sunni Arabs. Men uniformed and equipped as Iraqi government forces abducted Sunni Arabs in the Baghdad area, causing all Sunni Arabs to view government forces with suspicion.

Iraq is currently an unstable balance of conflicting groups. Kurdish separatists and Shi'ite Arab militias are represented in the government. Sunni Arab insurgents have long withstood U.S. forces and apparently feel confident of their ability to defy less-capable Iraqi government forces. The nascent government could sink into irrelevance as the country disintegrates into warring factions along ethnic and sectarian lines. Moreover, the dissolution of Iraq would tempt neighboring states to intervene, possibly leading to a regional crisis.

The conflict in Iraq involves unusual alignments. Former Ba'athists, who once held secular views, are aligned with foreign fighters, who hold extreme Islamic views. The United States, which had earlier supported Saddam Hussein against Iran, finds itself aligned with Shi'ite Arabs, who have close ties to Iran. On all sides are militias and irregular forces, ranging from the well-established Pesh Merga, which is in effect a national army, to Sunni Arab resistance organizations, which seldom rise much above the level of small armed groups. U.S. forces were initially an occupation force and subsequently became an ally of the Iraqi government, but they are now sometimes a neutral force between warring sects.

Kurdish Separatists

Kurds stay in Iraq as a matter of convenience, although they desire independence. Under the U.S. occupation, some Pesh Merga entered the New Iraqi Army, which was initially composed almost entirely of them and Shi'ite Arabs. Recognized as a legal militia, the Pesh Merga thrived and took on new life. The principal Kurdish movements have stayed at peace with each other, and their leaders hold high office in the Iraqi government.

Although secure within their homeland, the Kurds face dangerous issues on their borders. On their northern border, Turkey is alarmed by Kurdish separatists on its territory and aggrieved that the United States refuses to classify them as terrorists. Kurdish resistance groups continue to operate in small numbers across the Turkish-Iraqi border. To the south, the Kurds are determined to hold what they regard as key territory, especially the city of Kirkuk. In addition, the Kurds are threatened by the extended effects of growing anarchy within the rest of Iraq. Having previously suffered from Iraq's strength, the Kurds are now threatened by its weakness.

Although Kurdish separatism currently occasions less violence than the conflict between Sunni and Shi'te Arabs, it has threatening aspects. Shi'ite Arabs may not allow Kurds to maintain a state within a state without demanding similar concessions for themselves. If Shi'ite Arabs insist on comparable rights in southern Iraq, as advocated by SCIRI leaders, Iraq would cease to be a unified country. Moreover, if Sunni and Shi'ite Arabs eventually resolved their differences, the issue of Kurdish separatism might become more acute.

Sunni Arab Insurgents

The Central Intelligence Agency (CIA) defines insurgency as "a protracted political-military activity directed towards completely or partially controlling the resources of a country through the use of irregular military forces and illegal political organizations."[1] Insurgents want to control particular areas, in contrast to terrorists, who do not strive to create an alternative government. Soon after the invasion, Sunni Arabs perceived the United States to be siding with Shi'ite Arabs against them. From their perspective, they rebelled against a collaborationist regime imposed by a foreign occupier, not a legitimate government.

---

[1] Central Intelligence Agency, *Guide to Analysis of Insurgency*, quoted in Daniel Byman, *Going to War with the Allies You Have: Allies, Counterinsurgency, and the War on Terror* (Carlisle, Pa.: U.S. Army War College, Strategic Studies Institute, 2005).

During the first year after the invasion, the coalition estimated that there were roughly 5,000 active insurgents; it later revised its estimate to 20,000, while Iraqi government officials estimated much larger numbers. However, the number of insurgents at any one time was less important than the insurgency's ability to recruit new members. The pool of potential recruits may have comprised several hundred thousand Sunni Arabs, most of them young and unemployed.

The insurgents are predominantly Sunni Arabs living in central and west-central Iraq, i.e., the Sunni Triangle, including Baghdad. Prominent among them are former Ba'athist officials, including senior officers from the Iraqi armed forces and men from the Fedayeen Saddam and Ba'ath Party militia. Indeed, 99 of 200 generals who served in the old Iraqi Army were probably active in the insurgency during 2006.[2] In northern cities such as Kirkuk, where many Shi'ite Arab oil workers had settled, the Ba'athist regime developed a system of safe houses to help suppress any uprising, a system that survived the invasion. Saddam Hussein had initiated an Islamic Revival campaign prior to the invasion in an attempt to bolster his popularity. During the insurgency, searches of insurgent havens uncovered both Ba'athist and Islamic literature.

Sunni Arabs fear that Shi'ite Arabs and Kurds will unite against them and oppress them in revenge for past injuries. In addition, they fear loss of their Sunni identity in an Iraq they no longer dominate. Initially, Sunni Arabs boycotted elections, which they associated with the U.S. occupation and domination by Shi'ites. In the December 2005 national election, Sunni Arabs voted in large numbers and accepted office in the Iraqi government, but the insurgency continued unabated.

There is no single organization or umbrella group that speaks for the insurgents. Many of the organizations that issue public pronouncements appeal to Islamic sentiment, not Ba'athist ideology. Al-Moqawma al-Iraqiya al-Wataniya al-Islamiya (Iraqi National Islamic

---

[2] Interview with DIA/J-2, Chief, Iraqi Intelligence Analysis Branch, Arlington, Va., May 5, 2006. Former Chief of CJTF-7/C-2 Iraqi Analysis Branch, Baghdad.

Front) operates in the Sunni Triangle and may be a group comprising several organizations. Jaish Ansar al-Sunnah (Followers of the Sunni Army) operates in northern Iraq from Baghdad to Kurdish areas and proclaims a stridently Islamic ideology. Jaish Muhammad (Army of Muhammad) operates in the Sunni Triangle from Ramadi to Baquba and threatens to attack regional states that intervene in Iraqi affairs. Jaish al-Islami fil-Iraq (Islamic Army of Iraq) is composed of Salafists with ties to foreign extremists.[3]

U.S. officials negotiate with Sunni Arab insurgents[4] but are frustrated by their intransigence and incoherence. The insurgents have no publicly visible central leadership and no publicly declared goals beyond the departure of U.S. forces from Iraq. In contrast to the highly visible leadership of al Qaeda in Mesopotamia, the leaders of the Sunni insurgency remain largely anonymous. Due to this lack of public leadership, the United States and the Iraqi government find negotiation extremely difficult. The more-moderate Sunni insurgents have said that they would disarm after death squads were eliminated, Shi'ite militias were disarmed, amnesty was offered to Sunni Arab insurgents, and key political demands were met. Some of the Sunni insurgents seem to oppose al Qaeda in Mesopotamia and have even attacked its members.[5]

To prevail against U.S. forces, the insurgents do not have to win engagements; they merely have to survive and inflict losses. They operate in small bands equipped with light and some heavy infantry weapons. It is unclear whether the bands are organized at any higher level. They typically initiate contact when they choose and subsequently dis-

---

[3] Ahmed S. Hashim, *Insurgency and Counter-Insurgency in Iraq* (Ithaca, N.Y.: Cornell University Press, 2006, pp. 170–176); Anthony H. Cordesman, "Iraq's Evolving Insurgency and the Risk of Civil War" (working draft), Washington, D.C.: Center for Strategic and International Studies, May 24, 2006, pp. 146–150).

[4] Ambassador Zalmay Khalilzad, "The Next Six Months Will Be Critical" (Interview, *Der Spiegel*, June 7, 2006). As of October 1, 2007: http://www.csmonitor.com/2007/0301/p99s01-duts.html.

[5] Department of Defense, *Report to Congress in Accordance with the Department of Defense Appropriations Act 2006 (Section 9010): Measuring Stability and Security in Iraq* (Washington, D.C.: Government Printing Office, August 2006, p. 29).

appear into the civilian population. The insurgents have a large recruiting pool of embittered, unemployed youth, including demobilized soldiers and members of criminal gangs. Failure to reconstruct Iraq, especially to relieve the massive unemployment, has helped keep this recruiting pool large. Sunni Arabs tend to see insurgents as defending them against U.S. occupiers and Shi'ite Arabs. Where the population is sympathetic, or at least passive, insurgents have great freedom to act against U.S. forces without fear of exposure. In addition, their presence tends to influence people not to support an apparently powerless Iraqi government.

Insurgents are responsible for most of the bombings in Iraq, but terrorists achieve some of the most spectacular effects. Sunni Arab insurgents conduct most of the attacks on U.S. forces, typically through large roadside bombs. Extremists, many of them foreign fighters, usually conduct attacks on easier targets, such as government officials and Shi'ite Arab civilians. Most, if not all, of the suicide bombers appear to be foreign extremists who either come to Iraq prepared to die or are persuaded to conduct suicide attacks after their arrival. Suicide bombing in Iraq is unprecedented in its scale and the devastation inflicted, especially on Shi'ite Arab civilians. However, there are indications that regional approval of such terrorism is declining due to revulsion at the carnage inflicted on Muslims.[6]

The Sunni insurgents have survived because they could replace their losses and are genuinely popular among Sunni Arabs. They have a sufficiently large recruiting pool to replace their losses in combat. Most Sunni Arabs over time have come to see the insurgents as their defenders against the U.S. foreign occupiers and the Shi'ites.

Violent Extremists

Despite its public support for the Palestinian cause, the Ba'athist regime had little interest in terrorist groups and suppressed them within Iraq.

---

[6] See, for example, George Michael and Joseph Scolnick, "The Strategic Limits of Suicide Terrorism in Iraq" (*Small Wars and Insurgencies*, Vol. 17, June 2006, pp. 113–125).

It did not harbor al Qaeda, as the Taliban regime had done in Afghanistan, or Palestinian and Lebanese terrorists, as Iran does.[7] However, Saddam Hussein welcomed foreign help, especially from Ba'athist Syria, shortly before the invasion.

Terrorism impedes the reconstruction of Iraq. It tends to drive international agencies and NGOs out of the country, causes middle-class Iraqis to flee, prompts donor countries to reconsider making monetary contributions, discourages private companies from investing, compels the United States to divert funds toward security, and disrupts projects that are under way.

The most spectacular terrorist attacks are made against Shi'ite gatherings in every venue: pilgrimages, weddings, funerals, open-air markets, restaurants, and even mosques. These attacks prompt retribution from Shi'ite vigilantes and militias against Sunni Arabs, causing a spiral of increasing violence, especially in some Baghdad neighborhoods. As sectarian violence increases, other groups also perpetrate terrorist acts, especially in Baghdad. Death squads abduct people, torture them with electric drills, murder them, and leave their mutilated bodies in public places.

Coalition authorities believe that foreign extremists constitute only about 10 percent of the active fighters in Iraq. However, in June 2005, more than 1,100 of the 4,100 detainees in the Abu Ghraib prison were from other countries.[8] Foreign extremists have disproportionate influence on the conflict, because they are willing to sacrifice themselves in suicide bombings. They use tactics perfected during the struggle against Israel, especially suicide bombing, which they direct against U.S. forces, Iraqi government officials, and, increasingly, Shi'ite Arabs.

Some of the violent extremists are Salafists, but others exhibit such motivations as animosity toward the West, self-sacrifice, and per-

---

[7] For a review of U.S. intelligence on the Ba'athist regime vis-à-vis al Qaeda, see Select Committee on Intelligence, United States Senate, *Report on Postwar Findings About Iraq's WMD Programs and Links to Terrorism and How They Compare with Prewar Assessments* (Washington, D.C.: Government Printing Office, 2006, pp. 60–112).

[8] Col. James Brown, "Baghdad Correctional Facility" (speech delivered at the Baghdad Correctional Facility (Abu Ghraib), Baghdad, June 5, 2005).

sonal shame that others are fighting and dying in Iraq. Most are not terrorists with global aspirations who would attack the United States if there was no war in Iraq. On the contrary, they are motivated primarily by that war, which they perceive as occupation of an Arab country by a neocolonial power allied with Zionism. They are loosely allied with Sunni Arab insurgents, but this relationship is tense because of divergent goals and traditional Iraqi distrust of foreign influence.

Extremists come from many Muslim countries besides Iraq, but primarily from Egypt, Saudi Arabia, and Jordan. They are recruited through mosques and during the annual hadj (pilgrimage). The umbrella organization for Salafists in Iraq was initially Jamaat al-Tawhid wa'al Jihad (Unity and Jihad Group), led by Abu Musab al-Zarqawi. This same group was later called Tanzim Qa'idat Al-Jihad in Bilad al-Rafidayn (Al Qaeda in Mesopotamia) after al-Zarqawi's pledge of allegiance to Osama bin Laden.

Extremists and insurgents exploit the media, especially regional television networks, to magnify their importance, to capitalize on U.S. mistakes, and to attract recruits. They routinely use digital cameras to record their actions, especially attacks with IEDs, to make themselves appear formidable. The television network al Jazeera in Qatar became notorious for broadcasting pictures of these actions and running programs sympathetic to the insurgents, who were often portrayed as brave fighters against foreign occupation. Extremists also use Internet sites to spread propaganda, publicize their actions, and attract new recruits. They have disseminated footage of beheadings of hostages, gruesome spectacles that were intended to intimidate enemies but may have hurt the extremists' cause.

Foreign extremists have had effects out of proportion to their numbers, initially bringing in bomb-making skills, recruiting or exploiting a cadre of suicide bombers and using them as precision shock troops, and initiating a drumbeat of negative propaganda about the occupation. The foreigners were motivated by diverse agendas, some personal and some related to jihad—a diversity that coalition information-operations specialists largely missed as a key vulnerability.

Shi'ite Arab Militias

When U.S. and Iraqi government forces fail to protect them, Shi'ite Arabs turn to militias for security. They join small neighborhood militias, the long-established Badr Organization, and the rapidly growing Mahdi Army. The Badr Organization and the Mahdi Army are rival militias that keep an uneasy peace with each other while both infiltrate the Iraqi police. Many Sunni Arabs see the Shi'ite militias as outposts of Iranian influence.

During the Iran-Iraq War, SCIRI was founded in Iran to organize Shi'ite resistance to Saddam Hussein's regime. Iran's Revolutionary Guards provided training and equipment to SCIRI's military arm, then called the Badr Corps, which fought alongside Iranian units. After the U.S. invasion of Iraq, SCIRI emerged as the leading Shi'ite party in Iraq, while continuing to receive assistance from Iran. The leader of SCIRI is Ayatollah Abd al-Aziz al-Hakim, the brother of Ayatollah Mohammad Baqir al-Hakim, who was assassinated in August 2003 by a car bomb as he emerged from worship at the shrine of Imam Ali in Najaf. Despite his ties to Iran, al-Hakim cooperates with the coalition and plays a prominent role in the government, enabling members of the renamed Badr Organization to assume positions in the security apparatus. Iran supports the Badr Organization but hedges by supporting the Mahdi Army as well. Iranian agents provide assistance to Shi'ite militias, including cash and explosive devices.

The Mahdi Army, led by Muqtada al-Sadr, arose after the invasion of Iraq and soon overtook the longer-established Badr Organization. Al-Sadr is the youngest son of the Ayatollah Mohammed Sadeq al-Sadr, assassinated in 1999 by agents of the Ba'athist regime. The vast Shi'ite slum in northern Baghdad was named Sadr City after Mohammed Sadeq al-Sadr. The younger al-Sadr is fervently Islamic, strongly nationalistic, and bitterly opposed to the U.S. presence in Iraq, which is still construed as an occupation. Although a poor public speaker, al-Sadr has the moral stature of one who suffered under oppression by the Ba'athist regime.⁹ Under his leadership, the Mahdi Army became an organization comparable to Hezbullah, providing social services in addition to security. His supporters currently control the Ministries of Health, Agriculture, and Transportation, while the Facilities and Protection Service is a source of funding and jobs for the Mahdi Army. By mid-2006, the Mahdi Army may have had as many as 60,000 fighters.[10]

Criminal Gangs

Criminal elements have a heavy but underreported impact on the Iraqi government's ability to govern. Saddam Hussein released large numbers of violent prisoners from Iraqi jails during the invasion, in the expectation that they would create anarchy. These and other criminals continue to plague Iraq today. Criminal gangs trade in drugs, smuggle petroleum products and cars, deal in stolen antiquities, and conduct kidnappings for ransom money. They produce explosive devices and emplace them for cash. They provide facilitators and foot soldiers for the insurgency. In addition, criminal gangs collude with corrupt Iraqi government officials to divert oil revenues. From early on, U.S. analysts underestimated the carryover effect of forces at play in Saddam-era Iraq: A ruthless dictatorship had masked widespread corruption and interaction between military officers and criminal smuggling enterprises centered in Baghdad and among the many tribes inside Iraq's border regions. Many of these relationships remain intact today and, moreover, form the backbone of the insurgent enterprise faced by coalition forces.

Insurgent Use of Terrorism

All of the insurgent groups in Iraq have employed terrorism—murders, bombings in public places, car- and man-portable explosive devices.

---

[9] Interview with Kadhim Waeli, Iraqi analyst, Headquarters U.S. Army Intelligence Command (Springfield, Va., April 9, 2006).

[10] Baker and Hamilton, 2006, p. 5.

Many of the attacks have been filmed and quickly posted on global media such as the Internet. Unrelenting terrorism acts, mainly in the form of persistent car bombings, have had several important effects and appeared focused on Baghdad from the start. These persistent attacks—which U.S. and Iraqi forces and supporting research agencies never sufficiently focused on until almost too late—spurred sectarian violence, impeded reconstruction, and required U.S. forces to tie up resources. Though foreign extremists may have been the most publicized perpetrators of terrorism, with walk-up suicide and car bombings, other hostile forces engaged in terrorist acts, including kidnappings conducted by criminal gangs that sold their hostages to insurgents and executions conducted by Sunni insurgents and Shia radical militias. In the summer of 2004, terrorists began to attack religious targets, primarily Christian churches in the Karrada district of Baghdad. The most spectacular attacks were against Shi'ite gatherings at pilgrimages, weddings, funerals, markets, restaurants, and mosques. These attacks prompted retribution from Shi'ite vigilantes and militias, causing Sunni citizens who were sitting on the fence to seek protection from Sunni insurgents and foreign fighters, further legitimizing their activities.

By late 2003, insurgents had begun to direct precise roadside-bomb attacks against coalition troops countrywide, supplemented by a steady stream of foreign suicide bombers. By the summer of 2004, the insurgents began to lay "daisy chains" of roadside bombs (multiple, interconnected weapons) in more-precise attacks involving squad-level, enemy-harassing attacks when coalition first responders arrived on the scene. By July 2004, the insurgency had begun to shift its attacks to the Iraqi people, whose protection was largely ignored by coalition forces, and Iraqi Police and Army protection plans, particularly in Baghdad. The lack of focus and commitment toward developing an antidote to this powerful mechanism for undermining Iraqi tolerance for U.S. troops proved to be one of the major strategy oversights on the part of coalition forces in the entire Iraq war. Congress, national agencies, and senior military commanders must share blame for the tragic consequences that ensued.

Since their introduction in late 2003, IEDs have become the single largest cause of casualties of both security forces and civilians in Iraq.

These weapons have been produced in huge quantities, mostly because they are very easy and cheap to make, and they have proved to be very effective for the insurgents. It is likely that the IED will emerge from Iraq as a weapon of choice for insurgent and terrorist groups around the world. Surprisingly, well-funded research organizations and analytical efforts in Washington never focused on the major IED killer of Iraqis—the vehicle-borne IED (VBIED).

CHAPTER THREE

Counterinsurgency in Iraq

To prosecute counterinsurgency in Iraq successfully, the United States needs a comprehensive strategy that includes a range of efforts or lines of operation. All these efforts should be mutually reinforcing, implying that the United States must make all of them simultaneously, but the appropriate weight of effort varies over time and by region of the country. The challenge is to find the right balance in rapidly changing circumstances.

Organization and Recognition of the U.S. COIN Effort Is Slow to Unfold

DoD initially took the lead for all matters concerning Iraq. The Office of Reconstruction and Humanitarian Assistance (ORHA) and later the CPA were subordinate to the Secretary of Defense, although the CPA Administrator also reported directly to the President.

DoD created ORHA to provide humanitarian assistance following the invasion. It was headed by Lt. Gen. (USA, ret.) Jay Garner, who had previously commanded Operation PROVIDE COMFORT, which provided humanitarian assistance to displaced Kurds following the Persian Gulf War. ORHA prepared for several kinds of humanitarian emergencies that did not occur, but it was not designed to support large-scale reconstruction.

After the CPA was established, civilian and military authorities shared control at the country level. Ambassador Bremer was Administrator for Iraq, with full authority under international law and resolutions of the Security Council to govern the country. Lt. Gen. (USA) Ricardo Sanchez, commanding coalition forces, was formally tasked to support Bremer, but Bremer and Sanchez worked autonomously. Bremer focused on negotiations with and among the Iraqi factions to secure their agreement on broad policy issues. Sanchez focused on military actions to suppress a growing insurgency, including sweeps and raids to disrupt insurgent groups and kill or apprehend their members.

On June 23, 2004, John Negroponte became the U.S. Ambassador to Iraq, and DoD ceased being lead for all matters concerning the country. The State Department took responsibility for political matters and eventually oversaw reconstruction through the Iraqi Reconstruction Management Office (IRMO). After Negroponte was chosen to become the first Director of National Intelligence on February 17, 2005, the post of Ambassador to Iraq remained vacant until Ambassador Zalmay Khalilzad arrived in Baghdad on June 21, four months later. On June 15, 2004, General (USA) George W. Casey, Jr., became Commanding General, Multi-National Force–Iraq (MNF-I). While subordinate to the regional combatant commander, General Abizaid, Casey also reported directly to the Secretary of Defense. As a result of these arrangements, two channels were established: The ambassador reports through the State Department on political matters, and the commanding general (first Casey, later Petraeus) reports through DoD on military matters. The Joint Strategic Planning and Assessment Cell coordinates their actions. As a result of increasing congressional interest in the events in Iraq, specific congressional deadlines for progress reports also have been established. While the commanding general in Iraq still reports to the President via the Secretary of Defense, General Petraeus must provide periodic updates to Congress, the first of which he delivered in September 2007.

Traditional U.S. Military Forces May Need to Be Adjusted

Conventional U.S. forces may be overwhelmingly superior to the insurgents with regard to traditional combat, but an Iraqi landscape highlighted by suicide and roadside bombings, assassinations, and beheadings is far from traditional combat. During the first two years of the war, U.S. forces repeatedly entered cities in the Sunni Triangle to engage the insurgents. They prevailed in most *traditional* tactical engagements, but the insurgency continued unabated because the insurgents continued to regenerate and replace their losses from within Iraq and outside Iraq. Moreover, these tactical, kick-down-the-door operations often increased hostility to U.S. forces among Sunni Arabs, who saw themselves as oppressed by a foreign occupation. As the war has continued, U.S. and Iraqi government forces have strived to achieve more control over cities in the Sunni Triangle, as outlined in the problematic clear, hold, and build approach largely foisted on them by the U.S. State Department planners.[1]

Just before Bremer went to Iraq, he received a draft report from the RAND Corporation, which estimated that the United States might need as many as 500,000 troops to stabilize Iraq, based on historical precedents. Bremer was stunned by this analysis and sent a summary of the report to Secretary of Defense Rumsfeld, but he received no reply.[2] On May 17, 2004, shortly before departing Iraq, Bremer conferred with Lt. Gen. Sanchez about the conflict. He asked Sanchez what he would do with two more divisions, and Sanchez replied, "I'd control Baghdad."[3] The following day, Bremer sent a memorandum to Rumsfeld urging the deployment of two additional divisions. According to General Pace, who subsequently became Chairman of the Joint Chiefs of Staff, the U.S. commanders responsible for Iraq and the Joint Chiefs analyzed the recommendation and agreed that troop levels were adequate.[4] At its peak in January 2005, the coalition had 184,500

---

[1] National Security Council, *National Strategy for Victory in Iraq* (Washington, D.C.: National Security Council, November 2005, pp. 2, 8, 20–21); President George W. Bush, "Strategy for Victory in Iraq: Clear, Hold, and Build" (speech at the Renaissance Cleveland Hotel, Cleveland, Ohio, March 20, 2006). As of October 1, 2007: http://www.state.gov/p/nea/rls/rm/2006/63493.htm.

[2] Bremer, 2006, pp. 9–10.

[3] Ibid., p. 356.

[4] Ann Scott Tyson, "U.S. Studied Bremer's '04 Bid for More Troops" (*The Washington Post*, January 13, 2006, p. A17).

troops in Iraq, of which 160,000 were U.S. troops. There were also approximately 58,000 contracted personnel performing support functions and another 20,000 providing security for private employers, for a total of 262,500.[5] After a brief decline in U.S. troop strength near the end of 2006, by spring 2007, an increase in troop strength for a "surge" forecast new levels near 160,000 troops. Among the best-known and most important operations were those for control of Fallujah, Tal Afar, and Baghdad.

Fallujah

The U.S. Army's 82nd Airborne Division initially had responsibility for Fallujah, but the city's inhabitants were strongly opposed to the U.S. presence. To avoid further incidents, the division largely stayed out of the city, conceding control to the insurgents. In March 2004, the Ambassador 1st Marine Expeditionary Force (MEF) acquired responsibility for Fallujah and developed a plan to take control in stages. On March 31, insurgents there ambushed a small convoy escorted by four Blackwater security personnel. As described earlier, a frenzied mob set fire to the vehicles, dragged the charred corpses through the streets, and hung two of them from girders of a bridge. The U.S. ordered Lt. Gen. (USMC) James Conway, Commanding General, 1st MEF, to take control of the city quickly. At the same time, U.S. Army forces became engaged against the Mahdi Army in several places, including the holy city of Najaf. Sunni members of the Governing Council threatened to resign if no ceasefire was negotiated in Fallujah, while UN representative Lakhdar Brahimi was outraged by the offensive. Faced with the prospect of losing support from both the Governing Council and the UN, President Bush decided to stop the offensive, although senior officials realized that U.S. forces would eventually have to control the city.[6]

During summer 2004, insurgents prepared to defend Fallujah against another U.S. attack. Immediately prior to attacking into the

---

[5] John J. McGrath, *Boots on the Ground: Troop Density in Contingency Operations* (Fort Leavenworth, Kan.: Combat Studies Institute Press, 2006, pp. 132–135).

[6] Bremer, 2006, pp. 333–337.

city, U.S. forces dropped leaflets warning civilians to leave Fallujah but if they stayed, to remain in their houses and lie down on the floor holding the leaflets when Marines entered.[7] By the time the attack occurred in November 2004, most of Fallujah's residents had fled the city.

The 1st MEF, reinforced by two heavy Army battalions, advanced into Fallujah from the north, a direction unexpected by the defenders. The Marines encountered small groups of insurgents concealed in houses and apparently operating autonomously. The Marines proceeded through the city systematically, clearing some 15,000 to 20,000 buildings, one after another. When they encountered resistance, the Marines often had to destroy enemy strongpoints with tank main guns, mortars, artillery, and air attack. Although all fires were aimed at specific targets, the city suffered extensive damage. Once in control of the city, the Marines established a permanent presence and assisted Iraqi forces in taking over responsibility. People began returning to the city under a system of population control intended to prevent insurgents from regaining footholds.

Tal Afar

Tal Afar is a city of some 200,000 inhabitants, many of them Turkmen, located 40 miles from the Syrian border in Ninawa Province. During the U.S. occupation of Iraq, the city became a conduit for foreign extremists entering from Syria and also a safe haven for insurgents. The Iraqi police force disintegrated, and Iraqi Army forces stayed outside the city. Within Tal Afar, insurgents took over mosques and used them as command centers. They organized their forces into battalion-sized units with sections for mortars and snipers. In May 2005, the 3rd Armored Cavalry Regiment (ACR) assumed responsibility for Ninawa Province. As a heavy force with relatively few dismounted soldiers, it was not ideally suited to regain control of Tal Afar.

The ACR commander, Colonel (USA) H. R. McMaster, approached the problem of Tal Afar methodically. Ultimately he was

---

[7] Lt. Gen. (USMC) John F. Sattler, "Second Battle of Fallujah—Urban Operations in a New Kind of War" (interview with Patricia Slayden Hollis, *Field Artillery Magazine*, March–April 2006, pp. 4–8).

convinced by officers in the 5th Special Forces Group to move Iraqi Army forces out of static positions—the initial plan—and use them to conduct initial clear-and-sweep missions with Special Operations forces in some of the town's more troublesome districts. Aided by advisors from the 5th Special Forces Group, he conducted thorough reconnaissance of the area to uncover the power relationships.[8] The 3rd ACR was reinforced with the 2nd Squadron, 14th Cavalry, and it partnered with the 3rd Iraqi Army Division, an Iraqi Border Police Brigade, and Iraqi police throughout Ninevah Province. On Iraqi advice, the 3rd ACR constructed an eight-foot berm entirely around Tal Afar to control movement in and out of the city. From May though July 2005, the 3rd ACR concentrated on developing detailed intelligence of the area through reconnaissance and raids. It discovered that the enemy was organized into four battalion-sized units, and its stronghold was in the Sarai district, a maze of alleys and densely packed stone buildings, where armored vehicles could not deploy. The enemy had an effective low-level air defense, consisting of a network of observers and large volumes of small-arms and machine-gun fire.

In late August, the 3rd ACR began a methodical advance into the city, aided by a dismounted lead force of the 5th Special Forces Group accompanied by Iraqi Army units intent on isolating enemy forces in the Sarai district. As in Fallujah, U.S. forces allowed time for noncombatants to leave before the final attack began. Reinforced with the 2nd Battalion, 325th Airborne Infantry, 82nd Airborne Division, the U.S. force cleared the area, going house to house. The 3rd ACR also had support from two companies of Special Forces, one of which led the way through the most troubled neighborhoods on foot, an unusual concentration of their strength. Additional Iraqi Army and police forces moved into the city and established a permanent presence. These forces reformed the city administration, raised new police forces, and started reconstruction, including restoration of sanitation facilities, to effect an immediate improvement of living conditions. The

---

[8] Colonel (USA) H. R. McMaster, briefing, RAND's Washington Office, April 5, 2006; George Packer, "The Lesson of Tal Afar" (*The New Yorker*, April 10, 2006).

enemy responded with suicide-bombing attacks intended to intimidate the populace.

The 3rd ACR found that successful COIN required very close cooperation with Iraqi government forces at the tactical level and adopting at times unconventional suggestions from members of the 5th Special Forces Group. It established patrol bases where small numbers (less than a company) of U.S. and Iraqi government forces were permanently located. These patrol bases allowed U.S. and Iraqi forces to maintain close relations with the civilian population through continuous dismounted patrols and checkpoints. Every patrol was a combined force, i.e., included both U.S. and Iraqi personnel. The U.S. personnel treated their Iraqi counterparts as equals and trained them primarily by allowing them to observe how U.S. soldiers perform in combat. However, Special Forces officers continued to advise—even with much early resistance—that the 3rd ACR should have conducted combat-advising operations to simultaneously monitor and encourage nearby Iraqi Army units.[9]

Baghdad

Baghdad is a city of 6 million to 7 million people and the administrative center of Iraq. During the summer of 2004, Maj. Gen. (USA) Peter W. Chiarelli, Commanding General, 1st Cavalry Division, and Task Force Baghdad, was responsible for operations in the city. Chiarelli had to confront insurgency from Shi'ites living in Sadr City and key Shi'ite cities of southern Iraq led by Muqtada al-Sadr. Realizing that military confrontations between U.S. forces and insurgents would ultimately win more adherents to the insurgency, Chiarelli put Iraqi forces in the forefront to demonstrate that Iraqis were in charge of their country. At the same time, he placed greater emphasis on a "nonkinetic" approach to operations, providing economic works as a method of fos-

---

[9] John Gordon and Edward O'Connell, notes on visit to the 5th Special Forces Group, Fort Campbell, Ky., October 5–6, 2006.

tering good will for the coalition, while simultaneously putting city residents to work on projects to improve people's lives, such as repairing the sewage system, collecting garbage, and assuring the supply of potable water. From August to October 2004, Chiarelli's forces fought running battles with al-Sadr's Mahdi Army before negotiating a cease-fire, after which attacks declined dramatically and the residents of Sadr City began to benefit to some extent from the various projects. Though this apparent success was never proven by "cause and effect" analysis, Chiarelli concluded that such projects should be initiated more widely in Baghdad, but he lacked sufficient funds to implement them.[10]

Chiarelli believed reconstruction projects were fundamental to success, especially projects that employed large numbers of Iraqis and tangibly improved the community.[11] In his view, it was vital to give Iraqi citizens a sense that they would benefit from a peaceful future. Therefore, he fostered projects to collect garbage and repair the sewers that were flooding Sadr City with filth, projects that helped promote a lasting ceasefire with the Mahdi Army. However, Muqtada al-Sadr took some of the credit for these reconstruction efforts, and his movement still controls Sadr City. Some have argued that Chiarelli's approach may have had the opposite of the desired effect: The sudden infusion of cash from these projects may have created economic circumstances that *increased* the prospects for local militias by providing the area's unemployed youth with a bonanza of DVD players with which to watch extremist propaganda videos, cell phones to spot for insurgent cell leaders, and cars from which to drop IEDs by the side of the road.

By early 2005, the Multi-National Force–Iraq Strategic Communications Directorate was issuing edicts such as "Don't do anything to create more insurgents," which may have had the unintentional effect of inhibiting aggressive coalition security initiatives at all levels.

---

[10] Maj. Gen. (USA) Peter W. Chiarelli and Maj. (USA) Patrick R. Michaelis, "Winning the Peace: The Requirement for Full-Spectrum Operations" (*Military Review*, July–August 2005, p. 17).

[11] See Chiarelli, Maj. Gen. (USA) Peter W., "The 1st Cav in Baghdad, Counterinsurgency EBO [effects-based operations] in Dense Urban Terrain" (interview by Patricia Slayden Hollis, *Field Artillery Magazine*, September–October 2005, pp. 3–8).

In January 2006, the Multi-National Force–Iraq assessed the security situation in Baghdad and four of Iraq's 16 other provinces as "critical," implying "a security situation marked by high levels of AIF [anti-Iraqi forces] activity, assassination and extremism."[12] Small bands of men, some composed of al-Sadr's followers, roamed Baghdad at night, abducting and killing their opponents. The Baghdad morgue accepted 1,815 bodies during July 2006, of which about 90 percent had suffered violent death.[13] Some people began to relocate, moving into areas where their sects predominated. In June 2006, U.S. and Iraqi forces began a large-scale operation code-named "Forward Together," to improve security in Baghdad. The following month, the United States moved the 172nd Stryker Brigade from northern Iraq to Baghdad. On August 7, U.S. and Iraqi-government forces conducted a raid in the Shi'ite stronghold of Sadr City against individuals that a U.S. military spokesman described as "involved in punishment and torture cell activities." The Shi'ite Prime Minister, Nouri al-Maliki, expressed anger at the raid and promised, "This won't happen again."[14]

Air Support

Operations in Iraq focus on land forces and tend to demand large numbers of dismounted troops. However, air forces make important contributions, especially in reconnaissance and strike. Reconnaissance with unmanned aerial vehicles (UAVs) greatly improves commanders' views of their areas of operations. Strike, generally conducted as close air support, occurs less frequently in COIN than in conventional combat but still plays an important role, for example, in reducing insurgent strongholds.

All of the services employ UAVs for reconnaissance in Iraq. The most useful and heavily tasked is the Air Force's Predator, an unmanned, turbocharged aircraft equipped with synthetic-aperture radar and

---

[12] Multi-National Force–Iraq, *Provincial Stability Assessment Report* (January 31, 2006).

[13] Andy Mosher, "Baghdad Morgue Tallies 1,815 Bodies in July" (*The Washington Post*, August 10, 2006, p. 20).

[14] Andy Mosher, "U.S.-Backed Operation Targets Shi'ite Slum" (*The Washington Post*, August 8, 2006, p. 16).

electro-optical and infrared sensors that is flown remotely and can stay airborne for 24 hours. In addition to flying reconnaissance, Predators attack with laser-guided Hellfire missiles fitted with shaped-charge, blast-fragmentation, or augmented-charge warheads. The Army's principal UAV is the RQ-5A Hunter, a twin-boomed aircraft equipped with electro-optical and infrared sensors that is able to stay airborne for 18 hours. UAVs provide persistent surveillance of areas of interest, such as suspected terrorist safe houses, urban centers, lines of communication, and terrain surrounding major bases. When not armed, they provide data to other systems for rapid engagement of targets.

Prior to the invasion, the U.S. Air Force fielded a prototype of the remote-operations video-enhanced receiver (ROVER), a man-portable laptop computer that receives streaming data from airborne sensors. Equipped with ROVER, a joint tactical air controller can see pictures gained by sensors mounted on UAVs, fighters, and bombers. ROVER IV includes a point-and-click feature that allows the operator to designate a target on the display and send that designation to the attack aircraft. This is especially useful when the air controller and other observers do not have line-of-sight to the target and therefore cannot determine its location by lasing. Moreover, it eliminates the need for talk-on, i.e., verbal description of the target, which can be time-consuming.

Combating Improvised Explosive Devices

In Iraq, U.S. forces encounter huge numbers of IEDs, ranging from crude land mines to sophisticated command-activated devices. Against this threat, U.S. forces have fielded a wide variety of countermeasures, but IEDs still continue to take a steady toll of U.S. lives and inflict terrible wounds.

IEDs are the weapon of choice for insurgents and terrorists in the country, because they are extremely difficult to counter and have important effects. Roadside bombs, or remote-controlled IEDs (RCIEDs), are responsible for about half the casualties suffered by U.S. forces there,[15] while VBIEDs are responsible for the majority of Iraqi casual-

---

[15] Clay Wilson, *Improvised Explosive Devices in Iraq: Effects and Countermeasures* (Washington, D.C.: Congressional Research Service, 2005, p. 1).

ties. IEDs help recruit for the insurgency and hamper the movement of U.S. forces outside their fortified bases. They compel U.S. troops to keep civilian Iraqi vehicles at a distance and engage those who approach them too closely, thus driving a wedge between U.S. forces and Iraqi civilians. They increase the cost of war and divert U.S. effort toward countermeasures. IED attacks on August 19 and September 22, 2003, caused the UN to remove most of its staff from Baghdad.[16] In addition, attacks against Shi'ite Arab civilians ignite sectarian conflict.

The first IEDs were mortar and artillery shells with crude detonating devices. As the war continued, IEDs became more sophisticated, falling generally into three categories: command- or pressure-detonated roadside bombs, vehicular bombs, and explosive vests. Two roadside devices might be arrayed, for example, with the second explosion timed to kill people as they responded to the first. IEDs are often command-detonated, using electronic devices such as cell phones and garage-door openers, which enable bombers to select their targets. Vehicular bombs can be very large, often in the 1,000-lb class, and are usually delivered by suicide drivers. Many vehicular bombs are directed against civilian targets, such as public ceremonies and marketplaces, to inflict as many casualties as possible. The bomb builders have responded to countermeasures by using larger arrays of explosives, more-sophisticated command detonation, and larger numbers of suicide bombings.

To defend against IEDs, U.S. forces have hardened their administrative and combat vehicles and deployed specialized bomb-clearance vehicles. High-mobility, multipurpose wheeled vehicles (Humvees) initially received improvised armor plating and subsequently were fitted with uniformly produced armor kits. Stryker vehicles were equipped with additional plates and slat armor to defend against rocket-propelled grenades. The U.S. Army acquired the Buffalo, a commer-

---

[16] Independent Panel on the Safety and Security of the United Nations in Iraq, *Report of Independent Panel on the Safety and Security of the United Nations in Iraq* (New York: United Nations, 2003); The Open Society Institute and the United Nations Foundation, *Iraq in Transition, Post Conflict Challenges and Opportunities* (New York: Open Society Institute, 2004, p. 38).

cially produced, 23-ton, truck-like wheeled vehicle with armor protection and a remotely operated hydraulic arm, which was used to handle suspected devices. U.S. forces use a variety of small robots that are operated remotely and equipped with robotic arms and video cameras. They also employ Warlock jamming devices against some of the command links used to activate or detonate IEDs.

Aerostat-mounted systems were initially deployed to the theater to counter man-portable air-defense systems near runways and to defend the perimeters of some forward operating bases. However, it soon became apparent that they were useful against IEDs as well. Predators are also useful, but they are in short supply and are usually dedicated to other missions. U.S. forces employ snipers against persons emplacing IEDs, apparently with some limited success, but there are little data with which to evaluate their effectiveness. In general, offensive "left of boom" targeting measures such as snipers and Quick Reaction Forces (QRFs) were employed too late and with little effect against the roadside-bomb threat, while relatively little strategy or thought was devoted to the main killers of the Iraqis, the VBIEDs.

Detainee Operations

Currently, insurgents are classified as Iraqi citizens subject to civilian law. Therefore, they may be detained and subsequently incarcerated only when they are suspected and eventually convicted of offenses under rules of law. When U.S. personnel detain insurgents, they have to collect evidence of offenses chargeable under Iraqi law and thus act as though they were police.

Between 2003 and 2005, coalition forces detained approximately 50,000 suspected insurgents in Iraq. As of May 1, 2006, the coalition held 14,000 detainees in custody, but as of this writing, that number had increased to nearly 20,000.[17] The detainees included individuals from diverse circumstances and with quite diverse motives. Once an Iraqi citizen is detained, the first challenge is to discover why he joined

---

[17] The coalition tracks detainees who have been in custody longer than the two- to three-week intelligence hold period or have entered a brigade internment facility. It does not track those held for shorter periods of time.

or supported the insurgency. Knowing the detainee's motive is important not only for exploiting him, but also for devising strategy to defeat the insurgency. In the case of young detainees, important information can sometimes be garnered from their families. Unfortunately, for a period of more than four years, funding for a comprehensive motivation and morale study of the insurgency was rejected by authorities and agencies in Washington, D.C., as well as by some senior officers forward.

From September 20 through December 13, 2003, at least 24 serious incidents of abuse of prisoners occurred at the Abu Ghraib detention facility.[18] Maj. Gen. (USA) Antonio M. Taguba conducted an investigation and concluded that Military Police working at the prison were inadequately trained for their mission and were overwhelmed by conditions at the facility. An independent panel chaired by former Secretary of Defense James R. Schlesinger concluded that the abuses were acts of purposeless sadism related to a failure of military leadership and discipline. It found that "in Iraq, there was not only a failure to plan for a major insurgency, but also to quickly and adequately adapt to the insurgency that followed after major combat operations," for example, detainee operations.[19] Even in light of their growing importance for successful COIN operations, detainee operations received surprisingly little attention in the Army's new COIN Field Manual 3-24.

The detainee effort has serious flaws, including faulty collection of evidence, poorly assembled dossiers, lack of properly trained interrogators, and premature release of detainees. By 2005, the Iraqi Tribunal increasingly controlled disposition of detainees and set a high threshold of proof. Obtaining this proof requires collection of evidence, analysis of networks, and sophisticated interrogation. When the U.S. forces

---

[18] For a summary of these incidents, see Maj. Gen. George R. Fay, Investigating Officer, *Investigation of the Abu Ghraib Detention Facility and 205th Military Intelligence Brigade*, Baghdad, 2004.

[19] James R. Schlesinger et al., *Final Report of the Independent Panel to Review Department of Defense Detention Operations* (Washington, D.C.: Department of Defense, 2004, p. 10). The panel members were The Honorable Harold Brown, The Honorable Tillie K. Fowler, and General (USAF, ret.) Charles A. Horner.

fail to collect enough evidence to support indictments, dangerous individuals may be released. In addition, detainees are often released prematurely because the judicial system is overloaded and immature.

The United States failed to exploit detainees sufficiently for motivational insights, which could have proven very effective in calibrating kinetic and nonkinetic operations throughout the COIN effort. Nor has it until recently attempted to incite individual or group defections from their ranks to the government side. Some detainees are hardened insurgents; some are on the fringes of the insurgency; and some are innocent citizens apprehended through large, often indiscriminate sweep operations. Thorough processing would reveal the character of the detainees and could open possibilities for obtaining information and improving relations with the communities from which they come. At the brigade level, detainees may be held for two weeks, but most of them are interrogated only once, due to lack of interrogators. Moreover, many repeat detainees are hardened to the process and easily withstand interrogation.[20] Insurgents realize that detention is usually brief and followed by no repercussions, so they tend to regard it as an inconvenience. In some instances, detainees use detention as a way to meet other members of the insurgency and share experiences and technical knowledge, such as bomb-making.

For Arab audiences, the photographs taken at Abu Ghraib, endlessly copied and disseminated through the media and the Internet, confirm their worst apprehensions about Americans. The photographs depict acts of sexual sadism inflicted by American women on Arab men, acts that are especially humiliating in a culture that emphasizes masculinity and personal honor. Weak leadership, inadequate resources, confusing guidance, and tangled command relationships all contributed to the chaotic conditions at Abu Ghraib, where these abuses occurred.

---

[20] Interview, S-2 analysts from 10th Mountain Division, Baghdad, November 2005 (names withheld on request).

U.S. Development and Support of Iraqi Forces

Ultimately, Iraqi government forces must provide safety and security for the Iraqi population, but building such forces from scratch has proven difficult and time-consuming. It is no surprise that the U.S. Multi-National Security Transition Command–Iraq (MNSTC-I), led by senior Army officers, set up for this purpose, focused less on the Iraqi police than on the development of the Iraqi Army. This proved to be a fatal oversight.

The Iraqi Police

Under the Ba'athist regime, police forces were weak, corrupt, and poorly equipped. When the regime fell, only the traffic police remained intact, while the others receded into the population or even became involved in the insurgency.

After the invasion, a team from the U.S. Justice Department calculated that 6,600 police trainers were required nationwide. On June 2, 2003, Bremer approved the team's plan but lacked funds to implement it. In March 2004, the United States decided to provide 500 civilian police trainers through DynCorps, under contract with the State Department. Some of them did not go into the field because of security concerns, but even so, 20 DynCorps trainers died in Iraq.[21]

In September 2003, General Abizaid, Commanding General, U.S. Central Command, recommended to Bremer that the U.S. military assume responsibility for police training. Bremer opposed this recommendation, because he doubted the U.S. military's ability to train police and because he suspected that the military wanted to replace its own troops with poorly trained Iraqis.[22] Bremer saw his suspicion confirmed when Lt. Gen. Sanchez, Commanding General, Combined Joint Task Force-7 (CJTF-7), told a senatorial delegation in early October that 54,000 police were on duty, a number that Bremer found unbelievable. When he inquired about police training, he was told, "The Army is sweeping up half-educated men off the streets, running

---

[21] Michael Moss and David Rohde, "Misjudgments Marred U.S. Plans for Iraqi Police" (*The New York Times*, May 21, 2006).

[22] Bremer, 2006, pp. 168–169.

them through a three-week training course, arming them, and then calling them 'police.' It's a scandal, pure and simple."[23] After receiving this report, Bremer directed Sanchez to stop recruiting police.[24]

Despite Bremer's intervention, the U.S. military ultimately acquired responsibility for building an Iraqi police force. It tasked Military Police units and reservists with civilian police experience to serve as trainers and advisors to the Iraqi police. It coordinated these efforts through the Civilian Police Advisory Training Team (CPATT), an underresourced activity that subsequently came under the MNSTC-I. Efforts to develop the Iraqi police were not fully resourced, because the U.S. military initially devoted more attention to establishing a paramilitary force called the Iraqi Civil Defense Corps (ICDC), intended to fill the void caused by dissolving the Iraqi armed forces.[25] Moreover, MNSTC-I was focused on building the Iraqi Army and neglected to monitor the police adequately.

Initially, the Iraqi police received pistols, but subsequently they acquired AK-47 assault rifles and machine guns to match the firepower of the insurgents. Even with these weapons, many policemen stay in their police stations, afraid to run the risk of patrolling. Bombs are frequently detonated at their stations or in other areas where they congregate. The National Police are more effective, but the force is composed largely of Shi'ite Arabs. During early 2006, young Sunni males were repeatedly bound or handcuffed and killed by shots to the head. Death squads within the Iraqi police service, which had been infiltrated by militias, may have perpetrated some of these killings. Others might have been perpetrated by militia members wearing police uniforms, which are easily obtained from markets in Baghdad.

Border security is another difficult task. Approximately 30,000 border police are deployed on Iraq's borders in some 285 fortified stations. These stations were rebuilt and renovated to provide protection

---

[23] Statement by Doug Brand, Senior Advisor to the Ministry of the Interior, on October 3, 2003, recounted in Bremer, 2006, p. 183.

[24] Ibid., p. 186.

[25] Seth G. Jones et al., *Establishing Law and Order After Conflict* (Santa Monica, Calif.: RAND Corporation, MG-374-RC, 2005, p. 51).

during small-scale attacks by insurgents and criminal elements. In some areas, the border police are too few and too ill-equipped to maintain control of the border except on the most highly traveled routes. On the Syrian border in particular, they must contend with a legacy of smuggling that thrived during the Ba'athist regime to evade UN sanctions. U.S. military units advise and assist the border police, but there are not enough of them to control the border. As a result, insurgent and terrorist groups continue to receive new recruits and supplies from neighboring countries.

The Iraqi Armed Forces
Under Saddam Hussein, the Iraqi Army suppressed Kurdish and Shi'ite revolts with great brutality. In light of this history, Bremer and the CPA's Senior Advisor for National Security and Defense, Walter Slocombe, saw the Iraqi Army as an instrument of Sunni oppression of Kurds and Shi'ites. In Slocombe's words, the Iraqi Army was "a badly trained, ethnically unacceptable [i.e., Sunni-dominated] army with very dubious politically loyal leadership."[26] Slocombe also pointed out that the Iraqi Army had effectively disbanded itself by simply going home and that its facilities had been trashed by looters. In support of the decision to disband the Iraqi army, Bremer found that Shi'ite Arab leaders were strongly opposed to reconstituting it.[27]

Bremer formally dissolved the Ministry of Defense and a variety of security agencies, along with "the Army, Air Force, Navy, the Air Defense Force, and other regular military services."[28] Dissolution of the Iraqi armed forces caused discontent among former servicemen, who were suddenly unemployed. After weeks of demonstrations, the CPA agreed to pay ex-soldiers, but not to recall them. In fall 2003, the CPA began to establish a new Iraqi Army, originally envisioned as a 60,000-man force without responsibility for internal security.

---

[26] Walter Slocombe, interview, "Frontline" (Public Broadcasting System, August 17, 2004).

[27] Bremer, 2006, p. 55; see also pp. 235–236.

[28] Coalition Provisional Authority, *Order Number 2, Dissolution of Entities* (Baghdad, May 23, 2004, p. 4).

By late 2006, the Iraqi Army numbered about 138,000, but it displayed serious weaknesses. It was equipped almost entirely as light-infantry battalions supported by motor-transportation regiments. It included only one mechanized brigade, part of which was equipped with tanks and infantry fighting vehicles donated by Eastern European countries. As a result, there is a wide disparity between the lightly equipped Iraqi Army forces and the heavily equipped U.S. forces trying to accomplish similar missions. In addition, the Iraqi Army mirrors the sectarian and ethnic divisions that plague the country. Kurds, Sunni Arabs, and Shi'ite Arabs usually serve in battalions that consist largely or exclusively of their own groups. There is no judicial system within the Iraqi Army to assure discipline, and soldiers can refuse orders with impunity.[29]

Assessing Progress in Counterinsurgency

Progress in COIN is difficult to assess, because many factors are involved and data are often incomplete or missing. However, three indicators of progress are Iraqi casualties, reconstruction in critical sectors, and Iraqi public opinion.

Iraqi Casualties and Displacement

Neither the Iraqi government nor the U.S. government regularly releases data on Iraqi civilian deaths.[30] According to the United Nations Assistance Mission for Iraq (UNAMI), the Iraqi Ministry of Health recorded over 3,000 violent civilian deaths per month between July and October 2006, with the highest total, 3,709, occurring during October.[31] The causes of these deaths include terrorist acts, roadside bombs,

---

[29] Department of Defense, *Report to Congress in Accordance with the Department of Defense Appropriations Act 2006 (Section 9010): Measuring Stability and Security in Iraq* (Washington, D.C.: Government Printing Office, August 2006, p. 58).

[30] Hannah Fischer, *Iraqi Civilian Death Estimates* (Washington, D.C.: Congressional Research Service, RS22537, 2006, p. 1).

[31] United Nations Assistance Mission for Iraq (UNAMI), *Human Rights Report, 1 September–31 October 2006* (New York: United Nations, 2006, pp. 1–2).

drive-by shootings, kidnapping, military operations, police abuse, and incidents of crossfire. Because of this violence, many Iraqi civilians have sought refuge abroad or become internally displaced. The United Nations High Commissioner for Refugees (UNHCR) estimates that approximately 1.6 million Iraqis are refugees outside Iraq. In addition, approximately 400,000 became displaced within Iraq after the bombing of the Ali al-Hadi Mosque in February 2006. Figure 3.1 shows the numbers of internally displaced families through 2005 and newly displaced families in 2006. Most of the newly displaced families are located in central Iraq, where the population is predominantly Sunni Arab or mixed Sunni-Shi'ite Arab.

From the summer of 2004 onward, Iraqi civilians and police suffered increasingly from explosive devices, especially car bombs and suicide vests. Figure 3.2 shows the numbers of Iraqi deaths from various types of explosive devices between February 2003 and April 2006.

The Iraqi Economy

The CPA spent only about 10 percent of the U.S. funding for reconstruction before its tenure ended. Partly as a result of procedures for letting contracts, much of the CPA funding was directed toward large-scale projects to improve the Iraqi infrastructure.

In critical sectors such as oil production, electricity generation, and potable water, Iraq fell short of levels attained during the Ba'athist regime. One consequence of de-Ba'athification was the collapse of civil administration within Iraq. In February 2006, the Multi-National Force–Iraq assessed the economic situation in Baghdad as "developing slowly," and that in five of Iraq's other 16 provinces was assessed as "critical," meaning "an economy that does not have the infrastructure or government leadership to develop and is a significant contributor to instability."[32] Table 3.1 presents an overview of Iraqi reconstruction through December 2005.

Insurgents frequently attack reconstruction projects. Attacks on oil infrastructure, in particular, impact directly upon the Iraqi government, which depends on oil revenue to meet its operating expenses. Insurgents attack pipelines, pumping stations, and, less often, refineries.

---

[32] Multi-National Forces–Iraq, *Assessment*, January 31, 2006.

Figure 3.1
Internal Displacement in Iraq in 2006

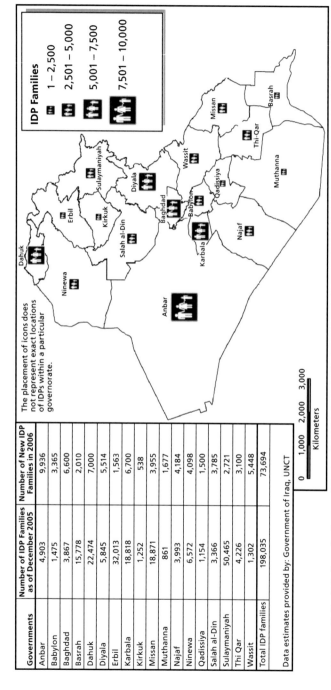

Governments	Number of IDP Families as of December 2005	Number of New IDP Families in 2006
Anbar	4,903	9,936
Babylon	1,475	3,365
Baghdad	3,867	6,600
Basrah	15,778	2,010
Dahuk	22,474	7,000
Diyala	5,845	5,514
Erbil	32,013	1,563
Karbala	18,818	6,700
Kirkuk	1,252	538
Missan	18,871	3,955
Muthanna	861	1,677
Najaf	3,993	4,184
Ninewa	6,572	4,098
Qadissiya	1,154	1,500
Salah al-Din	3,366	3,785
Sulaymaniyah	50,465	2,721
Thi Qar	4,226	3,100
Wassit	1,302	5,448
Total IDP families	198,035	73,694

Data estimates provided by: Government of Iraq, UNCT

SOURCE: United Nations Assistance Mission for Iraq (UNAMI), Movement of Internally Displaced Persons (IDPs) in Iraq as of October 2006, compiled by the Office of the United Nations High Commissioner for Human Rights. As of October 1, 2007: http://www.uniraq.org/maps/IDP%20movementOctober2006.pdf

RAND MG595/3-3.1

Figure 3.2
Iraqi Civilian and Police Deaths, by Cause

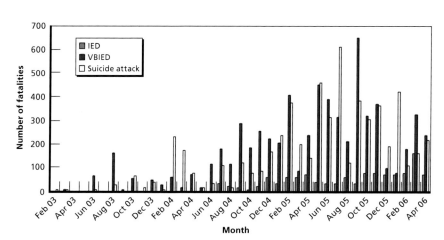

SOURCE: Data compiled by merging IraqBodyCount.org database and RAND–Memorial Institute for the Prevention of Terrorism (MIPT) Terrorism Incident database. As of October 1, 2007: http://www.iraqbodycount.org and http://www.tkb.org.
RAND MG595/3-3.2

Table 3.1
Iraqi Reconstruction Through December 2005

Category	Pre-Invasion	Post-Invasion Peak	December 2005
Power generation, MW	4,500	5,375	3,995
Electrical power in Iraq, hours/day	4–8	–	10.2
Electrical power in Baghdad, hours/day	16–24	–	3.7
Oil production, MBPD	2.58	2.67	2.0
Access to potable water, millions of people	12.9	8.25	8.25
Access to sewerage, millions of people	6.2	5	5

SOURCE: Special Inspector General for Iraq, "Summary IRRF [Iraqi Relief and Reconstruction Funds] Fact Sheet, Overview as of December 31, 2005." As of October 1, 2007: http://lugar.senate.gov/iraq/pdf/8_Bowen_Fact_Sheet.pdf.
NOTE: MBPD= million barrels per day; MW=megawatts.

From June 2004 through January 2006, insurgent attacks cost Iraq approximately $11 billion in lost revenue.[33] In addition, these attacks compelled the United States to divert a large portion of reconstruction monies—some 40 percent of the Iraqi Relief and Reconstruction Funds (IRRF)—to security.

The Iraqi government relies on the Facilities Protection Service and agreements with local sheikhs to protect key facilities. In addition, the Ministry of Defense contracted with British security firm Erinys International to train more than 14,400 guards to protect the oil infrastructure, but apparently it trained considerably fewer.[34] Traditionally, sheikhs receive small shares of oil revenue in return for protecting pipelines that run through their areas. But in some instances, sheikhs may have conducted their own attacks on pipelines to increase the perceived threat and hence their shares of revenue. In addition, looters frequently break into pipelines to steal oil and sell it at a higher price in foreign countries.

Iraq's extraction equipment is outdated and in poor repair, resulting in inefficient exploitation of the existing oil fields. The distribution system is also poorly maintained, and refining capacity does not meet even Iraq's own needs. The CPA tried to spur oil production but failed to meet its goal of 2.5 million barrels/day. By spring 2006, production was fluctuating at around 2 million barrels/day, about the preinvasion level. Even when oil prices rose steeply worldwide, Iraq still failed to generate sufficient revenue to fund reconstruction. This failure was due to government subsidies begun under the Ba'athist regime and to theft on a massive scale. The Special Inspector General for Iraq Reconstruction (SIGIR) testified that about one-third of Iraq's gasoline and diesel fuel was stolen each year.[35]

---

[33] Issam Jihad, statement reported by Kuwait News Agency, February 18, 2006, quoted by Onur Ozlu, *Iraqi Economic Reconstruction and Development* (Washington, D.C.: Center for Strategic and International Studies, 2006).

[34] Jonathan Finer, "U.S. Report Cites Progress, Shortfalls in Iraq Rebuilding" (*The Washington Post*, May 1, 2006, p. A9).

[35] Statement of Stuart W. Bowen, Jr., Special Inspector General for Iraq Reconstruction, "Iraq: Perceptions, Realities, and Cost to Complete," Hearing before the United States House of Representatives Committee on Government Reform, Subcommittee on National

The Ba'athist regime failed to modernize or even maintain the electrical power infrastructure, so that supply fell far short of demand. It masked this shortfall by serving key areas preferentially, especially the Baghdad area. After the invasion, generation of electrical power increased through the summer of 2005 but then declined, due in part to attacks on high-voltage lines and generators in southern Iraq. At the same time, Iraqis purchased large numbers of electrical appliances, almost doubling the demand for power. As a result, electrical power was available to Iraqi consumers nationwide less than half of each day during spring 2006.[36] In the summer of 2006, demand for electricity was expected to reach 10,000 megawatts, almost twice the available supply.[37] This shortfall was due not only to inadequate generation, but also to an antiquated and often nearly chaotic system of power distribution. In addition, Iraq suffers from looting of cables, destruction of transmission towers, and sabotage of pipelines.[38]

The State Department reports primarily on completed projects and the expected capability of treatment plants but provides little information on the quality of water reaching Iraqi households or their access to sanitation services.[39] In October 2003, the World Bank reported that none of Iraq's sewage-treatment plants were operational and that about half of the raw sewage was being discharged into rivers and waterways and subsequently used by Iraqis who lacked access to potable water. In summer 2006, a U.S. military officer working on water facilities

---

Security, Emerging Threats, and International Relations (Washington, D.C., October 18, 2005, p. 4).

[36] See U.S. Department of State, Bureau of Near Eastern Affairs, *Iraq Weekly Status Report*, April 19, 2006, p. 12. Nationwide average electricity availability was estimated at about 10.9 hours per day.

[37] Nelson Hernandez, "New Iraqi Power Feeds a Feeble Grid" (*The Washington Post*, May 1, 2006, p. A15).

[38] U.S. Agency for International Development (USAID), "Assistance for Iraq, Electricity" (Washington, D.C.: USAID, 2006). As of October 1, 2007: http://www1.usaid.gov/iraq/accomplishments/electricity.html.

[39] U.S. Government Accountability Office (GAO), *Rebuilding Iraq, Stabilization, Reconstruction, and Financing Challenges* (Washington, D.C.: U.S. Government Accountability Office, GAO-06-428T, February 8, 2006, p. 11).

estimated that water-treatment plants in Baghdad met only 60 percent of the city's needs. Garbage collection was also erratic or lacking, causing garbage to choke the city and clog sewer pipes, which then overflowed.[40] Only 9 percent of the urban population outside Baghdad was served by sewers, and rural areas had no piped sewage. As a result, much of the Iraqi population was at risk for water-borne diseases.[41]

Iraqi Opinion

More than three years into the conflict, most Iraqis suspect U.S. motives and think the United States is doing a poor job of reconstruction, yet many are willing to tolerate U.S. forces until Iraqi government forces become strong enough to assure security. Despite U.S. assurances to the contrary, most Iraqis think that the United States plans to have permanent bases in Iraq and would not withdraw if requested to do so by the Iraqi government.[42] In contrast, Kurds want U.S. forces to withdraw only as the security situation improves.

In January 2006, about half of all Iraqis approved of attacks on coalition forces, but attitudes varied by ethnic and sectarian groupings. Very few Kurds approved of such attacks, considerably more Shi'ite Arabs did, and the overwhelming majority of Sunni Arabs did.[43] It was obvious why Sunni Arabs would approve of attacks on coalition forces, but it was surprising that 41 percent of Shi'ite Arabs also approved. At the same time, Iraqis from all groups said that security would improve after U.S. troops withdrew and even that public services would improve.

---

[40] Anna Badkhen, "Violence Aside, Baghdad Is Broken" (*San Francisco Chronicle*, May 24, 2006, p. 1).

[41] Onur Ozlu, *Iraqi Economic Reconstruction and Development* (Washington, D.C.: Center for Strategic and International Studies, 2006, p. 30).

[42] In a poll conducted by KA Research Limited and D3 Systems, Inc., with respondents in all 18 provinces of Iraq on January 2–5, 2006, 67 percent of Kurds, 79 percent of Shi'ite Arabs, and 92 percent of Sunni Arabs said that the United States planned permanent bases in Iraq. Only 17 percent of Kurds, 32 percent of Shi'ite Arabs, and 5 percent of Sunni Arabs said that the United States would withdraw if it were told to do so by the Iraqi government.

[43] In the same poll, respondents were asked, "Do you approve or disapprove of attacks on U.S.-led forces in Iraq?" Sixteen percent of Kurds, 41 percent of Shi'ite Arabs, and 88 percent of Sunni Arabs expressed approval.

Most Iraqis said that the United States should be involved in reconstruction, but generally thought that it was doing a poor job.

In al Anbar Province, most Iraqis oppose the U.S. presence and the Iraqi government. In a recent poll, respondents said they felt threatened by the U.S. military, terrorists, criminals, militias, Arab fighters from outside Iraq, and the Iraqi Army. Depending on locality, they had differing views toward the Iraqi police. In Ramadi and western al Anbar Province, respondents said they felt threatened by the police, but in Fallujah, they saw the police as less threatening than the insurgents. When asked whom they supported, respondents put armed resistance and Iraqi police at the top of their lists, while Arab fighters from outside Iraq and U.S. military forces fell at the bottom. Table 3.2 presents responses for citizens in al Anbar Province, the heartland of Sunni Arab insurgency.

The first comprehensive poll taken in all provinces showed that Iraqis primarily trusted their families, friends, and Arabic television as sources of information. More important, the poll showed that young Sunni Arab males aged 18 to 24, the recruiting pool for the insurgency, were about three times as likely as the average Iraqi to trust what the insurgents said. They were twice as likely to trust what foreign fighters had to say, but still only 14 percent considered them trustworthy. Table 3.3 presents responses to the question, "Whom do you trust?"

Table 3.2
Degrees of Iraqi Support for Forces in al Anbar Province in 2006

Force	Fallujah	Ramadi	Western	Total
Armed resistance	3.97	4.20	4.33	4.17
Iraqi police in your district	4.13	1.34	2.08	2.97
Iraqi Army	2.12	1.35	1.50	1.67
Iraqi government	1.93	1.11	1.45	1.49
Arab fighters from outside Iraq	1.32	1.51	1.19	1.36
American Army [sic]	1.06	1.08	1.14	1.09

SOURCE: Lincoln Group, *Al Anbar Survey 7, 2006 Baseline* (Baghdad: Lincoln Group, May 2006, p. 14).

NOTE: Question: "Please tell me the degree of your support for the following forces." (1 = strongly not support, 5 = strongly support). The Lincoln Group noted that the surveys required the tacit approval of local authorities, some of whom were supporting the insurgency and some of whom were against it and therefore tended to skew the data.

Table 3.3
Iraqi Trust of Information Sources

Source of Information	Sunni Arab Males Aged 18–24	All Respondents
Family	99	89
Friends and neighbors	95	81
Non-Iraqi Arabic television	92	69
Moqawma (resistance)	62	24
Tribal leader (sheikh)	60	51
Foreign fighters	14	7

SOURCE: Lincoln Group Poll, "Iraqi Information Environment," March 2006.

NOTE: Respondents were asked, "Whom do you trust?" and were given a list of sources that did not include the coalition.

CHAPTER FOUR
Accounting for Success and Failure

The Bush Administration did not anticipate widespread, virulent resistance to U.S. occupation and to a new Iraqi government led by Shi'te Arabs and Kurds. As a result, it was initially unprepared to conduct COIN and promoted slow-paced creation of the government, ceding time to the insurgents. Ethnic and sectarian parties dominated the new government and failed to produce a foundation for national unity, as had occurred in Bosnia and Kosovo. Moreover, the government was so weak that even Shi'ite Arabs turned to militias for the protection that coalition forces failed to provide. In the absence of effective government, extremists in both sects committed outrages that made the division still harder to bridge.

Understanding Iraqi Society

Prior to invading Iraq, U.S. planners did not appreciate how Iraq's highly turbulent history would diminish the prospects for a new democratic order and planned instead for an unrealistic best case. Sunni Arabs had dominated the country during Ottoman rule, under British tutelage, and especially during Saddam Hussein's regime. U.S. planners should have anticipated that people accustomed to dominance would not willingly accept minority status in a new democracy. To supplement expertise within the U.S. government, planners should have turned to outside experts who were well acquainted with the tensions within Iraqi society.

In addition, the planners failed to understand how U.S. occupation, fundamentally military in character, would appear to Arab Iraqis of all sects. After World War II, the United States had replaced Britain and France as the principal external power in the Middle East. However unfairly, it was perceived as the inheritor of British and French colonialism. This perception was strong in Iraq, where Britain had exercised a mandate after the demise of the Ottoman Empire. Moreover, many Arabs in the Middle East perceive the United States as an uncritical supporter of Israel, which is seen as the relentless oppressor of Palestinian Arabs. Saddam Hussein deliberately inflamed this feeling among Iraqi citizens and presented himself as a champion of the Palestinian cause, for example, by firing ballistic missiles at Israel during the Persian Gulf War. Against this background, it was easy to predict that an occupation of Iraq would incite at least opprobrium, and very likely, resistance.

Little Planning for the Occupation of Iraq

Planning for the post-invasion period occurred primarily within DoD, which assumed a best case, i.e., no violent opposition, intact governmental apparatus, self-financed recovery, and rapid embrace of democratic practice. DoD made little use of studies done earlier by the State Department.

Planners structured U.S. forces for the invasion, which was much easier than anticipated, not for the subsequent occupation, which was much harder. Operations Plan 1003 called for some 500,000 troops, comparable to Operation Desert Storm in the Persian Gulf War, but Secretary of Defense Rumsfeld and General Tommy Franks incrementally reduced the number of troops by about 50 percent. When Army Chief of Staff General Eric K. Shinseki testified before Congress that "several hundred thousand" troops might be required after the invasion, Secretary Rumsfeld replied that his estimate was "far off the mark."[1] Failure to plan for potential resistance made it more likely to arise, since U.S. forces were not prepared to contain it. The forces were too small to effectively control even the Sunni Arab population.

U.S. planning would have been inadequate even had no resistance occurred. The planners should have at least envisioned how the United States would help Iraqis establish a democracy, since they had virtually no experience in democratic practice. Planning of that sort would have required an effort by all relevant departments of the U.S. government, not just DoD. Instead, DoD planners tended to assume that support would be forthcoming despite a dearth of planning. Had Iraqi governmental institutions survived the invasion, this lack of planning would have had less impact, but like other dictatorships built on fear, the Ba'athist regime collapsed when its forces were defeated. As a result, the Arab-inhabited parts of Iraq abruptly became ungoverned space, an eventuality the United States was unprepared to handle.

Planners also failed to address the problem of Iraq's porous borders and meddling neighbors. Iran was certain to take a strong interest in its co-religionists in Iraq and already had a conduit through SCIRI, which had been formed under Iranian sponsorship. Indeed, the United States had once supported Saddam Hussein against Iran out of apprehension that the Iranian revolution might extend to Shi'ite Arabs in Iraq. Syria was at odds with the United States because of its support of Hezbullah against Israel and was therefore also likely to oppose a U.S. occupation clandestinely. Despite these threats, the United States entered Iraq unprepared to secure its borders.

The Impact of a Lack of International Support for the War

In strong contrast to the 1991 Persian Gulf War, the invasion of Iraq in 2003 and the subsequent occupation lacked international support. Few foreign countries accepted the U.S. rationale for invading Iraq, i.e., that Saddam Hussein would provide weapons of mass destruction

---

[1] Reuters, "Shinseki Repeats Estimate of a Large Postwar Force" (*The Washington Post*, March 13, 2003, p. 12).

to terrorists, and the rationale eventually proved to be groundless. The war was initially unpopular in Islamic countries and became increasingly less popular. Even NATO allies, including France and Germany, expressed their opposition to the invasion. As a result, the United States could not obtain a UN Security Council resolution that would provide political cover for contributors.

Because the war was unpopular, coalition governments felt compelled to constrain the operations of their troops and to avoid casualties. The United States could expect few coalition forces other than UK forces to take an active role in combat. Instead, those forces stayed in well-protected bases and emerged only to conduct civic-action programs. Even at that, by late 2005, most of the major contributors had removed their forces or announced their intention to remove them.[2]

Even within friendly Arab states, media reporting usually opposed U.S. policy, especially on the highly emotional Palestinian issue. As a result, Arabs were predisposed to oppose the occupation of Iraq and to suspect U.S. motives. Most doubted that the United States had invaded Iraq to promote democracy, suspecting instead ulterior motives, such as securing access to Iraq's oil. In addition, Sunni Arabs strongly disapproved of U.S. support for Shi'ite Arabs in Iraq, fearing that they would promote a new Shi'ite power bloc allied with Iran. As the conflict continued, many Arabs in the region began to worry that the occupation and its aftermath were damaging the cause of democracy in the Arab world.[3]

The Disastrous Effects of Prematurely Dismantling the Ba'athist Regime

Under pressure from Shi'ite Arab and Kurdish leaders, Ambassador Bremer decided to abolish all Iraqi armed forces and institute a thor-

---

[2] For an overview, see Jeremy M. Sharp et al., *Post-War Iraq: A Table and Chronology of Foreign Contributions* (Washington, D.C.: Congressional Research Service, 2005).

[3] For an analysis based on opinion polling, see Shibley Telhami, "What Arab Public Opinion Thinks of U.S. Policy" (transcript of Proceedings, Washington, D.C.: The Saban Center for Middle East Policy, The Brookings Institution, 2005).

ough program of de-Ba'athification. These measures alarmed Sunni Arabs, who perceived them as an attempt to impose domination by Shi'ite Arabs.

Under Saddam, the Iraqi Army was led by Sunni Arab officers, while Shi'ite Arabs filled the lower ranks. However, the Army successfully defended the country against Shi'ite Iran and was one of the few national organizations that united the two sects. During the invasion, the Iraqi Army disintegrated through mass desertion, but soldiers could have been recalled to active duty, as U.S. military planners expected, or demobilized in ways that smoothed their reentry into civilian society and left them available for recall. Abrupt abolition of the Iraqi Army flooded Iraq with unemployed ex-soldiers who became recruiting pools for insurgent groups and militias.

During the Ba'athist regime, the upper echelons of most professions were pressured to become members of the Ba'ath Party. As a result, the party comprised practically the entire Sunni Arab elite, including educators and administrators, whose membership was little more than nominal. Sweeping de-Ba'athification not only disaffected Sunni Arabs, but also deprived the country of the services of many skilled professionals. A more selective program focused on Saddam's inner circle and his security apparatus would have dismantled Ba'athist rule without unnecessarily destabilizing the country.

The Challenge of Building a New Iraqi State from Scratch

DoD planners expected that a functioning government would remain after the invasion and could be reformed under new leadership. They expected that moderate, secular Iraqi leaders, including those drawn from the émigré community, could provide this leadership. Contrary to these expectations, the Ba'athist regime collapsed entirely, and émigré politicians had little popular following.

The CPA had full authority under the Geneva Conventions and a resolution of the UN Security Council to govern Iraq, but it lacked the power and perhaps also the will to govern. To fill the vacuum left by the vanished government, the United States initially established an occupation whose presence outside the Green Zone (the heavily secured area of Baghdad) was almost exclusively military. U.S. forces were highly visible, creating the impression of a heavy-handed military occupation that provided material for hostile propaganda.

The highly deliberate political process devised during the occupation took more than three years to produce the first democratically elected government under a new constitution. A proportional voting scheme was introduced, under which Iraqis voted overwhelmingly for parties that represented ethnic and sectarian factions. These parties had widely divergent views on Iraq's future and negotiated with each other for five months in early 2006 before forming a government.

Even after this long process, the resulting Iraqi government is ineffective, divided along sectarian lines, plagued by corruption, and absorbed with internal power struggles. Ministers put their offices at the disposal of their political parties and use public money to enrich their supporters. The Ministry of Defense is slow to support Iraqi Army units or even to assure that soldiers are paid promptly. Incompetence accounts for some failings, but in other cases, the ministries refuse to support units not controlled by their favored groups. The Ministry of the Interior has a particularly bad reputation, including complicity with Shi'ite Arab militias. Recognizing these failures, the Iraq Study Group recommended that the United States reduce support to the Iraq government if the government did not make substantial progress on a broad range of issues.[4] However, it is unclear whether the current government has either the will or the means to make such progress and become a true government of national unity.

Instituting a New System of Justice

Under Saddam, justice was perverted to perpetuate the regime, and prisons were used to torture political dissidents. The new Iraqi government must have a new system of justice to assure its legitimacy and help suppress violence, but the required prosecutors, judges, and

---

[4] Baker and Hamilton, 2006, p. 61.

confinement facilities are lacking. The U.S. Justice Department estimated that Iraq needed 1,500 judges, but by mid-2006, only 740 were serving. Moreover, judges are too intimidated to prosecute insurgents in some areas of the country.[5] Even more critical, there is no judicial system under which to impose discipline within the Iraqi armed forces, whose members can abandon their posts with impunity.

The Iraqi government does not maintain crime statistics nationwide or even in the key urban areas. In addition, Iraqis tend not to report crime because they lack confidence in the police.[6] Much violence is due to criminal activities such as theft, extortion, and kidnapping for ransom. Criminal activity and insurgency often overlap and complement each other, for example, when criminals are paid to emplace IEDs. Reliable crime statistics would help the Iraqi government and U.S. advisors to comprehend the scale of the problem and gauge success in law enforcement.

Another key shortfall is personal identification. Without identity documents that are biometric and tamper-proof, the Iraqi government cannot screen the population satisfactorily or monitor the movement of individuals. But the government has resisted introduction of a comprehensive system of identity documents, fearing that it would be unpopular.

Undertaking the Reconstruction of Iraq

During a counterinsurgency, reconstruction is not simply a humanitarian endeavor, like aid following an earthquake or a flood. Rather, it is an instrument the legitimate government can use to win allegiance of the people. The United States explicitly recognized this principle in its strategy of clear, hold, and build, where "build" implied reconstruc-

---

[5] U.S. Department of Defense, *Report to Congress in Accordance with the Department of Defense Appropriations Act 2006 (Section 9010): Measuring Stability and Security in Iraq* (Washington, D.C.: Government Printing Office, August 2006, pp. 10–11).

[6] Interview, Major Conrad Wiser, U.S. Embassy Public Affairs Military Information Support Team (PA-MIST), Baghdad, August 12, 2004. Wiser served as liaison to the new Iraqi police in the Baghdad area.

tion that would cement success by making people into stakeholders in a new and better system of governance.

U.S. planners assumed that Iraq could pay for its own reconstruction through oil revenue. When that assumption proved false, the United States began funding large contracts for improvement of major infrastructure such as power generation. These ambitious projects achieved disappointing results, largely because of deteriorating security. Most of the $18.3 billion earmarked for Iraqi reconstruction went to physical infrastructure, and none went to projects that would have alleviated unemployment among Iraqi youth. Lack of security also discouraged investment in Iraq. In late 2003, a few Iraqi émigrés began returning to Iraq, but deteriorating security subsequently caused many middle-class Iraqis to flee the country.[7]

In 2003, Deputy Secretary of Defense Paul Wolfowitz testified before Congress that Iraq would soon be able to finance its own reconstruction.[8] But in fact, the Iraqi government alone absorbed most of the country's oil revenue, leaving little for reconstruction. U.S. planners had not appreciated how decayed Iraq's oil industry had become and how much oil revenue was used to pay for subsidies of basic commodities. Iraq could not refine enough oil to cover its own needs and had to import refined products at considerable added cost. Worse yet, the Ba'athists had bolstered their popularity by massive subsidies of common consumer goods. Gasoline was priced far below its actual cost, while electricity was not even metered.

The United States was slow to establish a structure that could administer reconstruction aid at the provincial level. The State Department and DoD disagreed over responsibility for security of Provincial Reconstruction Teams (PRTs), with State insisting that DoD should take responsibility for securing them, while DoD argued that U.S.

---

[7] Interview, Bradford Higgins, former Director Joint Strategic Planning and Assessment Branch, U.S. Embassy Iraq, currently Assistant Secretary of State and Chief Financial Officer at the U.S. State Department, Washington, D.C., April 24, 2006; David Eders, "A Million Iraqis Flee War-Torn Country for Haven in Jordan" (*Washington Times*, May 27, 2006, p. 6).

[8] Paul Wolfowitz, Deputy Secretary of Defense, testimony before the House Defense Appropriations Subcommittee, Washington, D.C., March 27, 2003.

forces were overcommitted and should not take on new missions. In mid-April 2006, the State Department announced that U.S. forces would provide security for PRTs, but even then a DoD spokesman referred only to "facility and site security," leaving unclear whether movement would also be secured. By summer 2006, the United States had only four PRTs, located in Baghdad, Hillah, Kirkuk, and Mosul.

The Consequences of Failing to Maintain Security Early On

When the United States invaded Iraq, it did not anticipate an insurgency and was not prepared to counter one. The United States had not experienced a full-fledged insurgency against its own occupation since the Philippine Insurrection at the turn of the 20th century. The Vietnam experience was highly relevant but was largely ignored. Except within its Special Forces, the U.S. Army lacked doctrine for COIN and did not train its forces to conduct it. U.S. military commanders tended to see combat as their main mission and gave less attention to protecting Iraqi civilians.

Military Missions

From the beginning, there appeared to be too few U.S. forces in Iraq to secure the country. In spring 2003, in addition to an astounding lack of military direction in this regard, there were not even enough troops to stop Iraqi civilians from widespread looting of public offices and industrial facilities. For many law-abiding Iraqis who were in awe of U.S. capabilities during the invasion, this remains a watershed event in terms of disappointment, loss of trust, and the spiral downward. In fall 2004, insurgency broke out in Mosul, in part because the U.S. had replaced the 101st Airborne Division with a much smaller and less effective Stryker-brigade combat team. In summer 2006, the Marines lacked the strength in al Anbar Province to control cities such as Ramadi, where insurgents routinely engaged their patrols.[9] In late 2006, U.S. and Iraqi government forces were still trying to establish control of Baghdad, a task that demands large numbers of infantry and police forces. In addition, there were too few U.S. and Iraqi government forces to secure Iraq's borders, especially those with Iran and Syria.

When insurgency first became apparent, the United States responded with kick-down-the-door sweep operations that often netted few insurgents but made Sunni Arabs increasingly disaffected and created a mass of fodder overnight for a new wave of insurgent propaganda videos. There were never enough forces concentrated in troublesome cities such as Baghdad to provide security, and the U.S. military was too slow in making the security of the local population a primary mission, instead visibly concentrating on its own force protection, which appeared to locals to be at their expense. After sweeps and patrols pass through an area, control reverts to whoever previously displayed the greatest strength in these neighborhoods. As a result, though U.S. forces continue to conduct highly visible operations throughout much of Iraq, swaths of the country remain firmly under insurgent control.

The current U.S. clear, hold, and build approach calls for securing one locality at a time.[10] Under this approach, localities would gradually be brought under stable governance, and Iraq would be rebuilt from the bottom. But the United States and the Iraqi government have too few forces, too few resources, and too little basic agreement between them to implement this strategy. U.S. forces can clear any locality in Iraq, but they cannot hold more than a few localities at a time. Iraqi government forces should hold the cleared localities, but too often they lack either the capability or the will, in part because the U.S. program to train and equip Iraqi forces makes only slow progress. Few resources are available to build, because the United States has inadequate funds

---

[9] Ellen Knickmeyer, "U.S. Will Reinforce Troops in West Iraq" (*The Washington Post*, May 30, 2006, p. 1).

[10] See National Security Council, *National Strategy for Victory in Iraq* (Washington, D.C.: National Security Council, November 2005).

remaining, and potential donors are discouraged by the increasing violence.

Within some cities in the Sunni Triangle and Sadr City in Baghdad, civilians tend to see U.S. forces as threatening them and insurgent groups and militias as protecting them. As a result, they join these groups, contribute to their support, and provide cover for their activities. Very often, local imams incite the faithful to oppose the foreign occupiers, and mosques function as hide sites and organizing points or centers for resistance to the perceived occupation. This perspective on U.S. troops is not confined to Iraq. Indeed, U.S. forces are almost invariably described as occupiers in Arab media, evoking comparison with Israeli occupation of territory inhabited by Palestinian Arabs. In some areas of Iraq, the local population celebrates when U.S. forces are killed by explosive devices and applauds those who emplaced them.

Lack of Infiltration and Tips Hinders Intelligence on the Insurgency

The United States has had great difficulty acquiring human intelligence in Iraq. Disintegration of the Ba'athist police and intelligence services and a top-down strategy to replace them caused a major gap in this vitally needed capability. In addition, the United States lacks experts with regional expertise and the required linguistic skills.

Coalition counterintelligence experts spend much of their time vetting mid-level and senior Iraqi officials for government posts, rather than running networks of informers, a critical part of COIN. The United States is reluctant to approve a comprehensive amnesty program, which, if successful, could produce valuable intelligence.

In 2005, the coalition initiated a program called "Eyes on Iraq" to solicit tips from the populace. The coalition received many tips over time but kept no records on how many led to actionable intelligence. Lack of a professional, well-trained Iraqi police force also hampers collection of human intelligence. In addition, some Iraqi security forces are slow to grasp the proper role of intelligence. U.S. officials charged with training Iraqi security forces discovered that the Iraqis tended to view intelligence as a tool of state repression, which it was during the Ba'athist regime.[11] Additionally, until recently, counterintelligence efforts were blocked by senior officers not schooled in unconventional warfare, contributing to the enemy's almost unabated freedom of action at a neighborhood level.

---

[11] Interview, Colonel Calvin Wimbish, Chief MNF-I, Ministry of Defense, Intelligence Transition Team, U.S. Embassy, Baghdad, October 10, 2005.

CHAPTER FIVE
Building Effective Capabilities for Counterinsurgency

This chapter summarizes the capabilities required to conduct COIN successfully, based on experience in Iraq. The goal of COIN is to gain people's allegiance to the legitimate government, implying that they are confident the government and its allies—in this case, U.S. forces—will protect them and promote a future they desire. Signs of increasing allegiance include willingness to provide information on insurgents, take part in civic life, hold public office, serve as police, and fight as soldiers. The process is reciprocal: The government can better protect its citizens as they become more willing to fight for it. The art of COIN is to achieve synergy and balance among various civilian and military efforts or lines of operation (see Figure 5.1). The optimal balance usually varies by region and over time, so there is no single balance that would be universally appropriate.

Because COIN requires the harmonious use of civilian and military means, unity of effort is the *sine qua non* for success. Unity of effort implies that all relevant entities, military and civilian, are subject to a common control in pursuit of the same strategy. At the outset, the U.S. government needs an operational planning process that embraces all departments of government that can contribute to COIN. Currently, only DoD and to some extent USAID are accustomed to developing operational plans and executing them, and they cannot plan for the entire government. The planning process should originate at the highest level and extend throughout the government, as envisioned in Presidential Decision Directive 56. Through this process, the depart-

Figure 5.1
Illustrative Lines of Operation

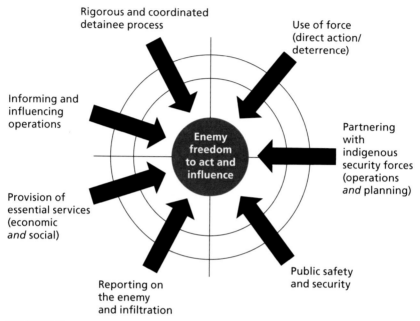

ments should develop a common plan, identify the implied tasks, assign responsibility to accomplish these tasks, and set standards for their accomplishment.

During execution of the plan, the interagency process should assure timely support from all departments and agencies of the government and provide a venue for devising strategy, under the direction of the President. It should resolve policy issues quickly, produce comprehensive instructions, and ensure that all departments implement those instructions. As illustrated by the PRT issue, the interagency process has not always provided timely support to operations in Iraq.

At the country level, there should be a unity of effort in all matters concerning the counterinsurgency, ideally through an individual vested with full authority. Separate channels for civilian and military efforts can function adequately if the leaders cooperate, but such cooperation is hostage to the vagaries of personality and to departmental politics. At the provincial level, a single individual should be authorized to integrate all U.S. actions, including those by conventional forces, Special Operations forces, and civilian agencies. That individual should be supported by a formal organization comparable to the PRTs currently operating in Afghanistan and Iraq. However, the appropriate model may vary, depending on circumstances.

Use of Force

Military force is one of several lines of operation that must be balanced in a successful COIN. It is a fundamental principle of COIN that military force alone can rarely bring success. This principle does not imply that military force cannot impose order if it is used freely and ruthlessly. Indeed, the Ba'athist regime imposed order primarily through the Iraqi Army, which was regularly employed to suppress unrest. But it does imply that a deeply rooted insurgency, one that enjoys widespread support from a population, cannot be defeated by military force alone. Even the Ba'athist regime needed a system of governance to maintain its power, albeit one based largely on intimidation, corruption, and manipulation of the divide between Sunni and Shi'ite Arabs. In any case, the United States does not have an option to use large conventional forces over an extended period against a deeply rooted insurgency. An effective long-term military occupation of Iraq against Sunni Arab opposition, for example, would require very large forces, probably larger than could be generated and sustained under the current U.S. all-volunteer system.

Counterinsurgency is frustrating for conventional-force commanders, especially if they have not trained for it. They are surrounded by unseen enemies who are often indistinguishable from innocent civilians. These enemies fight when they choose and break contact before conventional forces can overwhelm them. In Iraq, the insurgents open fire on U.S. patrols with infantry weapons, sometimes combined with IEDs. After their advantage of surprise has passed, they merge into their local neighborhoods. Their most effective weapon is the IED, emplaced where U.S. forces are likely to travel and often command-detonated. It would be much easier to fight such enemies if Iraqi civilians would inform on their activities. But too often civilians give information to U.S. and Iraqi authorities at great risk, and it is not acted on, reducing their willingness to provide tips in the future.

Even with the planned expansion of Army Special Forces, there will be too few to meet all demands, and therefore their efforts must be carefully prioritized. They should be employed as little as possible for tasks that conventional Army and Marine forces could perform. They should specialize in training indigenous forces and conducting combined operations with those forces. They should spend most of their time and resources engaged in combined operations, preferably with indigenous forces they have helped equip and train. Unless Special Forces are employed in this way, much of the leverage provided by their unique skills may be lost.

In Iraq, Special Forces are usually collocated with a conventional-forces commander at a forward operating base and work within his area of operations. As a result, their actions must harmonize with conventional-force operations in the same area. However, because they have separate chains of command, harmony often depends upon personal relationships. One possible strategy might be to assign Special Forces to larger conventional forces, but conventional-force commanders might lack the specialized knowledge necessary to employ them appropriately. In a better arrangement, Special Forces might be ordered to support conventional-force commanders under conditions specified in the order establishing this relationship. These conditions would help assure that Special Forces were employed appropriately and allowed scope for initiative. In addition, the Army should conduct training and exercises prior to deployment, to educate conventional-force commanders in special operations.

During COIN, it is normally preferable to employ U.S. military forces in combined operations with indigenous forces. Combined operations give the indigenous forces opportunities to emulate the skill and professionalism of U.S. forces. The Special Forces approach of "combat advising" is instructive in this regard. When U.S. forces operate alone or with token participation by indigenous forces, they are more likely to appear as foreign occupiers and be hampered by the language barrier and lack of familiarity with local conditions. However, in some situations, U.S. forces may have to operate alone. Immediately after an invasion, the U.S. military may be the only effective force in the country and therefore *de facto* responsible for all aspects of public order. Even later, indigenous forces may be unequal to the challenge, and U.S. military forces must accomplish tasks that are inherently difficult for foreign troops, such as protecting the population.

Combat in COIN often centers on engagements at tactical levels employing infantry weapons, but in Iraq the IED became the main form of tactical engagement for the insurgency. The forces in contact are often quite small, typically at squad and platoon levels. In urban terrain, the battlefield may shrink to a few city blocks or even a few houses where insurgents have taken cover. Combat tends to be inconclusive and fought as much for political and propaganda effect as for military reasons. For example, insurgents may seek engagements to demonstrate their patriotism and attract new members, even though they never prevail.

Insurgents usually have less firepower than counterinsurgent forces, but they can compensate through expedients. For example, they can employ indirect-fire weapons, such as mortars and rockets, and explosive devices of various sorts, such as land mines and car or truck bombs. To escape counterbattery fire, they can employ indirect weapons very briefly or from areas where they expect that U.S. forces will not be allowed to fire. In the absence of sophisticated range-finding, indirect fire may be so inaccurate that it seldom rises above the level of harassment. In contrast, explosive devices can be extremely effective and hard to counter. In Iraq, insurgents countered even the heaviest tanks—which had proven so effective in the initial invasion—by employing very large explosive charges and sophisticated warheads.

For conventional forces, greater firepower includes a wider range of indirect-fire weapons and air-delivered ordnance. This firepower is needed to protect convoys, suppress enemy indirect-fire weapons such as rockets and mortars, reduce insurgent strongholds, and quickly engage time-sensitive targets, such as enemy leaders. The United States is currently developing small bombs to fill the gap between guided missiles such as Maverick and general-purpose bombs that traditionally weigh

500 pounds or more. The United States is also improving its capability to laser-designate targets and communicate target coordinates digitally. Fighter aircraft are a very expensive way to provide constant coverage. Gunships are more efficient platforms, but during daylight, they are vulnerable to low-level air defense. One promising alternative would be a new platform that has most of the characteristics of a gunship but employs missiles rather than guns. Another solution, already being implemented with the multiple-launch rocket system (MLRS), is the use of the Global Positioning System (GPS) to guide indirect fire.

Table 5.1 lists capabilities associated with the use of military force and some of the implied tasks.

To accomplish these tasks, U.S. and indigenous forces normally require a mix of force types ranging from very light Special Operations forces to heavy forces equipped with large armored vehicles. Counterinsurgency is primarily an infantryman's war, and the emphasis is naturally on the availability of infantry, although mobility is also important. Tanks are still useful because they survive most of the weapons commonly available to insurgents and deliver very precise fire, typically controlled by the supported infantry. Infantry fighting vehicles transport infantry in relative safety from all weapons except large explosive devices, carry much-needed supplies, and provide fire support during contact.

Because of scarcity of forces and competing demands, the United States may have to leave some tasks almost entirely to indigenous forces. For example, border control is largely an Iraqi responsibility, although U.S. forces partner with some indigenous units and provide surveillance. Security of key facilities is also left to Iraqi forces, including local militias that contract to secure petroleum pipelines.

Public Safety and Security

The primary responsibility of any government is to protect its people. Indeed, no government can appear fully sovereign unless it fulfills this responsibility. Foreign forces can temporarily substitute for indigenous forces, but they are inherently less effective, and they may diminish the government's legitimacy.

Table 5.1
Use of Military Force and Some Implied Tasks

Capability	Implied Tasks
Protect civilian population (see also Table 5.2)	Know local culture and society (U.S. forces) Overcome the language barrier (U.S. forces) Develop sources of local information Conduct dismounted patrols Operate traffic-control points Secure centers of communal life Protect civilian leaders
Eliminate terrorists	Develop picture of terrorists Kill or capture key terrorists Disrupt terrorists' sources of funding and support
Suppress insurgents	Develop intelligence on insurgent groups Control weapons and munitions Clear urban areas Isolate and search areas of insurgent activity Respond quickly to contact with insurgent forces Conduct raids on concentrations of insurgents
Mitigate sectarian violence	Promote dialogue between and interaction of sects Protect religious observances and celebrations Control riots and suppress public disorder Monitor performance of indigenous security forces
Control borders	Keep border areas under surveillance Identify illicit traffic Interdict illicit traffic
Secure key facilities	Secure oil pipelines and refineries Secure airports and seaports Secure government centers
Protect U.S. forces	Reduce risks to dismounted forces Reduce risks to vehicular traffic Secure bases of operation Secure forward airfields

Securing a population requires sufficient numbers of police and other security forces, including military forces. When necessary, the United States employs military forces for internal security, as it did, for example, to enforce federal court orders during the civil-rights movement, to restore order in Los Angeles during the Rodney King riots, and most recently, to assure order in New Orleans following Hurricane Katrina. These cases illustrate the general concept that force should be abundantly available and sparingly used. The idea is to demonstrate overwhelming force, but to apply it in a restrained, disciplined way. During the Rodney King riots in California, for example, large numbers of National Guard troops deployed into the Los Angeles area and had a calming effect. The worst approach is to introduce forces that are large enough to incite opprobrium but too small to impose order.

Table 5.2 lists capabilities associated with security of the population and some of the implied tasks.

One way to reduce popular animosity is to minimize the risk to civilians from military actions. Improved procedures at checkpoints, for example, can reduce the risk of inadvertent engagement of innocent civilians. Even with safeguards, however, military forces will always pose some risk to civilians, especially during encounters with insurgents. In a firefight, it is rare to see enemy combatants. Normally, soldiers return fire against areas from which they believe incoming fire originated, not against individual targets, but eyewitnesses may interpret such fire as indiscriminate. Another way to reduce animosity is to inform civilians of military actions and advise them on how they can best avoid danger to themselves.

Partnering with and Enabling Indigenous Forces

Indigenous forces are central to successful COIN, especially in view of the fact that the ultimate goal is allegiance to the legitimate government. Eventually, indigenous government forces must bear the entire responsibility for protecting citizens and defending the country. If they enjoy a good reputation, such forces have strong advantages, especially

Table 5.2
Security of the Population and Some Implied Tasks

Capability	Implied Tasks
Reduce criminality	Elicit crime tips from the civilian population
	Promote comprehensive crime-reporting
	Build competent civilian police forces
	Establish a reliable criminal-justice system
	Establish systems of neighborhood policing
	Develop a national system of personal identification
Assure public order	Protect public gatherings
	Disperse and control riots
	Stop sectarian violence
Defend against terrorists and insurgents	Secure communal centers against attack
	Reduce risk of bombing attacks
	Suppress death squads
	Disband and demobilize militias
	Engage and destroy insurgent groups
Minimize risk from military action	Minimize collateral damage
	Prevent inadvertent engagement of civilians
	Inform the public of military operations
	Remove unexploded ordnance and land mines

at the local police level. They naturally enjoy local support, provided they are not identified with elements that people fear or despise.

Building indigenous forces can be a frustrating endeavor. It can be extremely difficult, even impossible, to recruit forces from a local population that is sympathetic to an insurgency or intimidated by insurgents—for example, through threats to family members. But forces from other populations may have scarcely more contact with the population than U.S. forces and may commit acts that inflame the insurgency.

To prevent abuses and encourage professionalism, U.S. forces should partner with indigenous forces down to the tactical level. Successful partnering is based on personal acquaintance and mutual trust, which are disturbed each time U.S. military units rotate. With each rotation, the newly arrived troops must get to know their Iraqi counterparts and earn their trust, while becoming familiar with their area of operations. Overlapping assignments can ease transition, but rotation still tends to reduce effectiveness, at least temporarily. As an alternative, the United States might rotate smaller elements at a time in order to preserve continuity. Table 5.3 lists capabilities associated with enabling indigenous forces and some of the implied tasks.

In Iraq, perhaps the greatest gap in capability would be in building police forces. U.S. civilian police forces have insufficient excess capability to generate the numbers of trainers and advisors required during a large-scale counterinsurgency. Even if enough civilian police were available, few would accept the risk of operating in areas of high insurgent activity. Insurgents recognize police as being particularly dangerous to their designs and make them priority tar-

Table 5.3
Enabling Indigenous Forces and Some Implied Tasks

Capability	Implied Tasks
Build police forces	Vet policemen from the former regime Recruit new members for the police force Train police in basic and advanced skills Equip police appropriately for their duties Assure logistical and administrative support Partner with indigenous police forces
Build military forces	Recruit new members for the military forces Train in basic soldiering skills Train in advanced specialties Equip with the appropriate range of weapons Assure logistical and administrative support Partner with indigenous military forces
Build border guards	Recruit new members for the border guards Train and equip border guards Provide permanent facilities along the border Share intelligence gained through technical means Partner with border guards
Promote ministerial competence	Monitor the performance of government ministries Provide advisor support Assure a reasonable level of competence

gets. Under such circumstances, protecting the police becomes a task that can consume already scarce military resources.

There are two readily apparent solutions to this problem, but neither is completely satisfactory. One solution would be to train indigenous police in neighboring countries, where they would at least be safe during the training process. However, they would also need to be mentored in their home stations, or their formal training might produce meager results. Another solution would be to use military police to train indigenous police and partner with them. In the U.S. system, most of the required military police units are in the Army National Guard. Some of the National Guard personnel are policemen and policewomen in their civilian lives and are highly skilled in civil law enforcement. Others are trained simply as military police and have little experience in civilian law enforcement. An ideal force for COIN would be trained both in military police duties and in civilian law enforcement. It would combine expertise in law enforcement with ability to fight as light infantry. The Italian Arma dei Carabinieri (literally, corps equipped with carbines), modeled on the French gendarmerie, fits this description. The Carabinieri are part of Italy's armed forces but execute police functions.

Reporting on the Enemy and Infiltration

During conventional war, U.S. forces require intelligence on enemy forces, including their organization, strength, deployment, and morale. Because morale is often considered the least important part of the assessment, U.S. forces may try to break the enemy's morale but not let success hinge on its breaking. Military commanders plan to destroy enemy forces, if necessary, making their morale relatively unimportant.

However, during COIN, priorities for reporting on the enemy are much different. It is still important to understand the organization, strength, and deployment of insurgent forces, but this information may be highly volatile and amorphous. Insurgents may change in the blink of an eye into apparently innocent civilians and back into insurgents again. Moreover, insurgent organizations may be highly decentralized and informal, to the point that even the insurgents have only a vague idea of their strength. But motivation, the equivalent of military morale, is a key factor. Insurgent forces and the civilians who support them are usually volunteers, although they may be pressed into service. Insurgents may intimidate people, but they do not have the apparatus of a state to compel obedience. Instead, they may appeal to motives such as political ideology, patriotic fervor, or religious faith, or they may resort to terror. For COIN, it is vital to understand these motivations so that they can be exploited and so that as many insurgents as possible can ultimately be co-opted.

During conventional war, intelligence is a top-down process until contact with the enemy starts to generate useful reporting. Typically, intelligence agencies at high command levels prepare orders of battle for enemy forces and communicate this intelligence to subordinate commanders. But in COIN, intelligence on the enemy is predominantly a bottom-up process based on reports from the field. In a successful COIN effort, much useful reporting is generated through police in daily contact with their neighborhoods.

During COIN, intelligence organizations at various levels usually infiltrate the opposing side and obtain inside information. When a population is of divided allegiance, there are usually opportunities for infiltration. Almost certainly, Sunni insurgents gain inside information from the Iraqi government and the Iraqi Army. Similarly, U.S. and Iraqi intelligence organizations cultivate informers among the insurgents. The process is not unlike that employed by law-enforcement agencies to penetrate criminal conspiracies such as the Mafia. Typically, a law-enforcement agency approaches members of the conspiracy who can be threatened, enticed, or cajoled into becoming informers, often with the promise of reward or immunity from prosecution. Table 5.4 lists capabilities associated with intelligence and some of the implied tasks.

Communications outside U.S. channels present considerable difficulties. The United States traditionally shares intelligence most readily with Britain, Canada, and Australia, and fairly easily with the members of the North Atlantic Treaty Organization (NATO); but beyond these security partners, sharing becomes increasingly problematic. During a counterinsurgency, it is axiomatic that the indigenous government is

Table 5.4
Intelligence on the Enemy and Some Implied Tasks

Capability	Implied Tasks
Understand the terrorist threat	Exploit technical intelligence on terrorists Elicit tips on terrorists Infiltrate terrorist groups Understand motivations of terrorists Learn how terrorists are recruited and trained Analyze the dynamics of terrorist networks Discern the terrorist trade craft Develop predictions of terrorist attacks Identify key nodes and personalities
Assemble a picture of the insurgency	Gain information from the civilian population Infiltrate insurgent groups Understand what motivates the insurgents Assess how to undermine motivations Analyze the dynamics of insurgent groups Discern evolving tactics of insurgent groups Develop models of insurgent activity Develop metrics for progress in COIN
Assess potential for sectarian violence	Assemble information on sectarian divides Study records of past conflicts Conduct public-opinion polls and surveys Understand what motivates people to become violent Assess how motivations could be defused Predict patterns of sectarian violence Track groups promoting sectarian violence
Communicate reports	Keep channels open for bottom-up reporting Disseminate reports quickly among U.S. forces Share reports with coalition forces Share reports with indigenous forces

infiltrated by the insurgents. The Iraqi government includes members of the Mahdi Army, which opposes the U.S. presence in Iraq and has fought U.S. forces on several occasions. Moreover, at least some of its Sunni members are very likely in contact with insurgent groups. Thus, the United States has to assume that information shared with officials of the Iraqi government will ultimately reach the insurgents, although sources and methods may be protected. Nevertheless, the United States still must share information to successfully support the Iraqi government. The aim is to share the information required for a successful partnership without giving too much information to the insurgents and to act on intelligence before the insurgents realize they are endangered.

Provision of Essential Services

There is no direct correlation between the well-being of a population and its propensity to support insurgency. People living in abject poverty may remain loyal to their government while more-affluent people rebel against it. However, people will expect government to provide essential services and will be at least disappointed by a government that fails to fulfill this responsibility. Moreover, reconstruction can be an important tool in COIN, especially when it is associated in people's minds with the legitimate government. In Iraq, people tend to blame the United States for their poor standard of living—for example, the dearth of electrical power. They even argue that the United States is deliberately withholding aid and inciting sectarian violence to perpetuate its stay in Iraq. This conspiracy theory seems absurd to Americans, but it reflects the disappointment of people who initially believed that the United States would assure prosperity.

During a counterinsurgency, provision of essential services and reconstruction should serve the overall purpose of gaining allegiance to the government. It follows that rebuilding projects should employ as much local labor as possible, even if the performance of such labor falls below international norms. Unemployed young men, especially discharged soldiers, are a primary recruiting pool for terrorist and insurgent groups, as well as militias. Moreover, projects should be designed to reflect credit on the government, not on an occupying power or on local militias. Above the minimum to sustain life, rebuilding projects should be predicated not simply on need, but rather on gaining people's allegiance. For example, they should be designed to complement and perpetuate successes in clearing areas of insurgents. Table 5.5 lists

Table 5.5
Essential Services and Reconstruction and Some Implied Tasks

Capability	Implied Tasks
Safeguard public health	Repair and maintain sewage system Assure reliable waste management Assure access to potable water Eliminate biohazards Assure adequate nutrition Provide ambulance service and hospitals
Assure public safety	Provide fire-fighting services Provide emergency medical care Remove unexploded ordnance safely
Reduce unemployment	Provide job retraining Offer large-scale employment on public projects
Promote economic recovery	Expand revenue from natural resources Increase power generation and transmission Restore and expand transportation systems Restore and expand communications systems Regenerate agriculture Promote a healthy climate for investment

capabilities associated with the provision of essential services and some of the implied tasks.

Most of the tasks associated with essential services and reconstruction require efforts by civilian departments of the U.S. government, especially the Department of State, USAID, the Treasury Department, and the Department of Justice. In addition, they require support from the World Bank, agencies of the UN, the European Union, and NGOs, some of which are supported through government channels. Unlike insurgent and extremist social-service providers, to make these disparate efforts useful in COIN, the United States must synchronize them at the country and provincial levels, not from Washington. A long-term, broadly based, countrywide scheme might be appropriate for development in a peaceful setting, but not during an insurgency. To conduct counterinsurgency successfully, the United States needs the ability to start and stop projects flexibly as part of a strategy to draw people into supporting their government, not by simply injecting money, but through personal interaction.

Informing and Influencing Operations

Official organs of the U.S. government have little credibility in the Arab world. It follows that the United States must make its case through Arab media such as al Jazeera, despite their bias against U.S. policy. The United States must rely primarily on factual reporting, because any attempt to slant the news will be immediately rejected. It should make stronger efforts to keep the Iraqi public informed, without attempting to portray the situation too optimistically. Table 5.6 lists capabilities associated with information operations and some of the implied tasks in Iraq.

A central problem for U.S. information efforts, and for U.S. policy generally, concerns commitment to democracy and human rights. U.S. policy seeks to further democracy in Iraq to provide an inspiration to people throughout the Arab world. This policy accords not only

Table 5.6
Information Operations and Some Implied Tasks

Capability	Implied Tasks
Inform Iraqi citizens	Convince Iraqis that the United States has no designs on Iraq Publicize civic action and reconstruction projects Explain the intent and scope of military operations Provide quick, accurate reports on current events
Counter opposing propaganda	Refute opposition claims to be "holy warriors" Adduce the suffering of innocent civilians Discredit al Qaeda as a foreign presence
Appeal to Arab publics	Escape from portrayal as an occupier Avoid identification with Israeli occupation forces Champion the cause of democracy and human rights
Build international support	Emphasize shared values Show respect for international norms Stress palatable alternatives

with U.S. values, but also with the aspirations of many Arabs, especially those who appreciate Western democratic practice. However, execution of this policy can be problematic. The United States has been reluctant to accept electoral outcomes that do not accord with its perceived interests. But if the U.S. commitment to democracy is to be credible, the United States has to accept outcomes that run counter to its wishes. Some friendly states that are highly important to U.S. policy are not democratically governed, in particular, Egypt and Saudi Arabia. Democracy in these countries would very likely produce governments less willing to cooperate closely with the United States than the present governments are.

Rigorous and Coordinated Detainee Operations

Detainee operations can assist COIN by removing dangerous persons from the society, generating sources of useful information, and in some cases, co-opting people who might otherwise have remained opponents of the government. In most cases, Iraqi detainees are processed through a civilian justice system, implying that there must be sufficient evidence against them to justify prosecution. Therefore, U.S. troops and indigenous forces must be skilled in gathering physical evidence and taking statements that prove the commission of crimes. After their arrest, Iraqi detainees should be held under humane conditions that demonstrate commitment to human rights. Torture and degrading treatment of detainees make the United States appear hypocritical and undermine the credibility of its policy. In the case of Iraq, perverse treatment by guards in the Abu Ghraib detention facility still colors regional perceptions of the United States. Once in detention, detainees should be thoroughly interrogated by personnel skilled in proper techniques and should be segregated into groups—potentially by motivational type—with hard-core terrorists isolated from the rest. Confinement facilities should not become schools for insurgency or meeting places for insurgents. Table 5.7 lists capabilities associated with detainee operations and some of the implied tasks.

Table 5.7
Detainee Operations and Some Implied Tasks

Capability	Implied Tasks
Detain suspected persons	Secure physical evidence and take statements Confine detainees under humane conditions Inform relatives of detainees' status
Process detainees	Segregate detainees by type Interrogate detainees methodically Make interrogations useful for all-source analysis Assemble complete dossiers on detainees Maintain a comprehensive database of detainees Move detainees through an effective justice system
Prevent recidivism	Identify detainees who might be "turned" Implement a program for reentry into society Track detainees after release

Some detainees may be willing to abandon the insurgent cause and reenter normal society. There should be programs to encourage such decisions and support reentry under decent conditions. In addition, U.S. and Iraqi government intelligence should track former detainees after their release to reduce the risk of recidivism.

CHAPTER SIX

Recommendations

The recommendations presented in this chapter are intended to assist the U.S. government in developing capabilities to conduct COIN. Counterinsurgency is a political-military effort that requires both good governance and military action. It follows that the entire U.S. government should conduct the effort.

Development of Strategy

The United States needs to improve its ability to develop strategy and to modify that strategy as events unfold. Strategy implies a vision of how to attain high-level policy objectives, employing U.S. resources and those of its allies. It also implies reflection upon strategies that adversaries might develop and how to counter them. Strategy should be developed at the highest level of government, by the President, his closest advisors, and his Cabinet officials, with advice from the Director of National Intelligence and regional experts, the Chairman of the Joint Chiefs of Staff, and the unified commanders. The unified commanders should link the strategic level of war with the operational level at which campaign planning is accomplished. A short, successful effort—e.g., the 1991 Persian Gulf War—may demand only the initial strategy. But a protracted, problematic effort, such as the current Iraq war, may demand modification of the initial strategy, or even a new strategy.

Coalition-Building

A high priority should be placed on building coalitions with like-minded countries to conduct COIN. Within NATO, development of capabilities to conduct COIN should be promoted, along with the projected agreement on creation of a special-operations division at the November 2006 summit meeting in Riga, Latvia.

Planning Process

A planning process should be developed that embraces all departments of the U.S. government. In the context of a national strategy, an office with directive authority should assign responsibilities to the various departments, assess their plans to discharge these responsibilities, request changes as appropriate, and promulgate a political-military plan that has enough operational detail to serve as an initial basis for execution of a COIN campaign. Presidential Decision Directive 56, intended to support planning for complex emergencies, outlines much of this planning process, although at a lower level of military action. To implement this directive, the United States needs to understand the history, culture, and nature of the society where combat operations will take place. It should have personnel proficient in local languages and knowledgeable in relationships among key elements of the society, in addition to employing outside experts.

In any future COIN operation, the United States should strive to quickly develop a coherent and balanced COIN strategy. The United States did not have a clear COIN strategy or plan in Iraq for more than three years. Senior military commanders and planners must establish an adequate mechanism with which to constantly assess performance in COIN operations. Senior military commanders must adapt, adjust, and/or modify strategy and tactics to meet the ever-changing demands of such operations. Commanders must closely monitor changing trends on the battlefield. In Iraq, senior military commanders were slow to understand and adapt to the changes in the enemy's strategy and tactics.

Unity of Effort

Unity of effort should be assured at country level and provincial levels encompassing all activities of the U.S. government, civilian and military. At country level, one individual should be given authority to direct all aspects of the U.S. effort. Depending on the situation, this individual might be either civilian or military. If he or she is a civilian official, the President should direct the highest-ranking military officer to provide support. If a military officer is given this responsibility, the President should direct the highest-ranking civilian to provide support. This individual should have authority over all actions concerning the country in question. He or she should represent the President in all aspects of U.S. policy regarding COIN within the country and related issues in the surrounding region.

Interagency Process

Because authority in Iraq is bifurcated between civilian and military authorities, unity of effort depends on the interagency process. Unfortunately, this process is not always sufficiently responsive, as illustrated by the PRT issue. The interagency process in Washington should be put on a wartime footing to conduct any COIN requiring large-scale U.S. forces. This process should support the person appointed by the President to prosecute the campaign within the parameters of the national strategy. The interagency process should be structured and operated to fulfill requests quickly and effectively. In addition, it must make changes to policy as necessitated by circumstances and the course of events. It is necessary to put a stop to "stove-piping," i.e., parallel but unconnected efforts of various departments and agencies. Senior military commanders and civilian officials should be given authority to reprogram funds without approval from Washington.

Host-Nation Governance

Preparations should be made to support governance in a host nation following the disintegration or collapse of a regime. Failure to quickly reestablish governance presents opportunities for insurgents and other groups opposed to the new government. Ideally, the civilian departments and agencies of the U.S. government should be prepared to provide advisors and technical personnel at short notice. Alternatively, the U.S. Army's Civil Affairs units could be expanded and resourced to fulfill this requirement.

Funding Mechanisms

The United States should be prepared to flexibly fund the establishment of foreign governments, the development of foreign militaries, and the reconstruction of foreign countries. Funding mechanisms should assure that funds may be moved flexibly across accounts, expended quickly in response to local contingencies, and monitored effectively by a robust, deployable accounting system. Civilian and military organizations should be prepared to award and monitor contracts with local companies to support rapidly changing situations. Senior military commanders must continuously reexamine the allocation of existing resources (both men and materiel) and make sure procurement priorities are in line with changing threats on the battlefield.

Counterinsurgency as a Mission

Counterinsurgency should be made a primary mission for U.S. military forces, on the same level as large-scale force-on-force combat operations. Military forces should train and exercise for interaction with civilian populations and insurgents in complex and ambiguous situations. Joint and service doctrine should treat COIN as a distinct type of political-military operation requiring far closer integration with civilian efforts than would be necessary for large-scale force-on-force combat operations. Counterinsurgency should be taught in staff colleges and other centers of advanced military education. Officers and senior non-commissioned officers (NCOs) should become familiar with foreign cultures and foreign military forces through education abroad, assignment to foreign militaries, and combined exercises.

Protection of the Indigenous Population

In future COIN operations, the United States must focus on security of the population as a critical measure of effectiveness. For too long, population security was not a priority in Iraq. Exceptional efforts must be taken to remove primary threats to the civilian populace. In Iraq, senior military commanders focused too long on roadside bombs and their impact on U.S. forces, rather than on the safety of the civilian population.

Personnel Policy

Personnel policies should be revised to assure retention of skilled personnel in the host country in positions that demand close personal interaction with the indigenous people. Legislation should be developed to enhance the quality and length of service of U.S. civilian personnel in the host country—in effect, a civilian counterpart to the Goldwater-Nichols reform.

U.S. Army Special Forces

Army Special Forces should be allowed to focus on training and operating together with their indigenous counterparts. Command arrangements should assure that Special Forces harmonize with the overall effort, while allowing scope for initiative. Special Forces might be directed to support conventional-force commanders under conditions designed to assure that they would be employed appropriately. In addition, the Army should conduct training and exercises prior to deployment to educate conventional-force commanders in special operations, especially those involving unconventional warfare.

Partnership with Indigenous Forces

U.S. conventional military units should be prepared to partner with corresponding units of indigenous forces. Partnership should imply continuous association on and off the battlefield, not simply combined operations. It should imply that U.S. military units adapt flexibly to conditions and mentor their counterparts in ways appropriate to their culture and skill levels. In COIN, U.S. success is inseparable from success of indigenous forces, which must ultimately assume the entire responsibility for security. U.S. Army Special Forces should assist conventional military units to prepare for partnership.

Policing Functions

The United States should prepare to conduct police work abroad and build foreign police forces on a large scale. DoD or other agencies in close coordination with it should prepare to introduce large police forces rapidly into areas where governmental authority has deteriorated or collapsed. These police forces should be trained and equipped to defeat enemies who are armed with weapons commonly employed by insurgents, such as crew-served weapons, rocket-propelled grenade launchers, and IEDs. They should be trained to partner with foreign police forces at every level, from street patrols to administration at the ministerial level. To improve handling of detainees, the United States should sponsor a large-scale study of detainee motivations.

Brigade Organization

Brigade-sized formations should be trained to conduct joint and combined COIN operations autonomously. These formations should include Army brigade combat teams and Marine Corps regimental combat teams that are designed to respond quickly to changing local conditions. They should include all the capabilities required by military forces to conduct COIN, including human-intelligence teams, surveillance systems, translators, and engineer assets. They should have the capability to field joint terminal-attack controllers down to company level during combat operations. They should be able to obtain intelligence support directly from national assets.

Gunship-Like Capability

The United States should develop survivable air platforms with gunship-like characteristics, i.e., comparable to those of the current AC-130 aircraft, to support COIN operations. These characteristics should include long endurance, fine-grained sensing under all light conditions, precise engagement with ordnance suitable for point targets, and robust communications with terminal-attack controllers. Moreover, these platforms should be survivable during daylight against low- and mid-level air-defense weapons, especially man-portable air-defense missiles, a capability lacking in the current AC-130 aircraft.

Intelligence Collection and Sharing

The ability to collect and share intelligence with coalition partners and indigenous forces should be developed. Special attention should be devoted to collection of human intelligence, including development of linguistic skills, interrogation techniques, and informant networks. It is essential to establish procedures and means to share intelligence rapidly with non-U.S. recipients at various levels of initial classification, without compromise of sources and methods.

Bibliography

"Al-Qaida Starts Own Online News Bulletin." *Kuwait Times* (Monday, October 10, 2005).

Badkhen, Anna. "Violence Aside, Baghdad Is Broken." *San Francisco Chronicle* (May 24, 2006): p. 1.

Baker, James A. III, and Lee H. Hamilton, Co-Chairs. *The Iraq Study Group Report*. New York: Vintage Books, 2006.

Bensahel, Nora, Andrew Rathmell, Thomas Sullivan, and Edward O'Connell. *Associating Iraqis with the C-IED Fight*. Santa Monica, Calif.: RAND Corporation, December 2005. Government publication; not releasable to the general public.

Black, Bob. Interview. "Controller, Persistent Threat Detection System." Camp Slayer, Iraq, September 15, 2005.

Bowen, Stuart W., Jr., Special Inspector General for Iraq Reconstruction. "Iraq: Perceptions, Realities, and Cost to Complete." Hearing before the United States House of Representatives Committee on Government Reform, Subcommittee on National Security, Emerging Threats, and International Relations. Washington, D.C., October 18, 2005, p. 4.

Brand, Doug, Senior Advisor to the Ministry of the Interior. Statement on October 3, 2003, in L. Paul Bremer III, with Malcolm McConnell, *My Year in Iraq, The Struggle to Build a Future of Hope*. New York: Simon & Schuster, 2006, p. 183.

Bremer, L. Paul III, with Malcolm McConnell. *My Year in Iraq, The Struggle to Build a Future of Hope*. New York: Simon & Schuster, 2006.

Burns, John F. "After a Long Hunt, U.S. Bombs the Leader of Al Qaeda In Iraq." *The New York Times* (June 9, 2006): p. 1.

Bush, President George W. State of the Union Address. Washington, D.C., January 28, 2003. As of October 1, 2007:
http://www.whitehouse.gov/news/releases/2003/01/print/20030128-19.html

———. Speech at the National Endowment for Democracy, United States Chamber of Commerce. Washington, D.C., November 6, 2003, p. 5. As of October 1, 2007: http://www.whitehouse.gov/news/releases/2003/11/print/20031106-2.html

Byman, Daniel. *Going to War with the Allies You Have: Allies, Counterinsurgency, and the War on Terror.* Carlisle, Pa.: U.S. Army War College, Strategic Studies Institute, 2005.

Celeski, Colonel (ret) Joseph. "Operationalizing COIN." Hurlburt Field, Fla.: Joint Special Operations University, Report 05-2, September 2005.

Chairman, Joint Chiefs of Staff. *Joint Tactics, Techniques, and Procedures for Foreign Internal Defense.* Washington, D.C.: Joint Chiefs of Staff, Joint Publication 3-07.1, April 30, 2004.

———. *National Military Strategic Plan for the War on Terrorism.* Washington, D.C.: Joint Chiefs of Staff, February 1, 2006.

Chiarelli, Maj. Gen. (USA) Peter W. "The 1st Cav in Baghdad, Counterinsurgency EBO [Effects Based Operations] in Dense Urban Terrain." Interview by Patricia Slayden Hollis, *Field Artillery Magazine* (September–October 2005): pp. 3–8.

Chiarelli, Maj. Gen. (USA) Peter W., and Maj. (USA) Patrick R. Michaelis. "Winning the Peace, The Requirement for Full-Spectrum Operations." *Military Review* (July–August, 2005): pp. 4–17.

Chirac, President Jacques. Televised address on March 18, 2003, reported by Elaine Sciolino. "Chirac Denounces Bush's Ultimatum, but Shows Willingness to Offer Some Help." *The New York Times* (March 19, 2003).

Coalition Provisional Authority. *Order Number 2, Dissolution of Entities.* Baghdad, May 23, 2004, p. 4.

Cordesman, Anthony H. "The Impact of the Iraqi Election: A Working Analysis." Washington, D.C.: Center for Strategic and International Studies, working draft, December 21, 2005.

———. "Iraq's Evolving Insurgency: The Nature of Attacks and Patterns and Cycles in the Conflict." Washington, D.C.: Center for Strategic and International studies, working draft, February 2, 2006.

———. "Winning the 'Long War' in Iraq: What the U.S. Can and Cannot Do." Washington, D.C.: Center for Strategic and International Studies, (working draft), April 3, 2006.

———. "American Strategic, Tactical, and Other Mistakes in Iraq: A Litany of Errors." Washington, D.C.: Center for Strategic and International Studies, working draft, April 19, 2006.

———. "Iraq's Evolving Insurgency and the Risk of Civil War." Washington, D.C.: Center for Strategic and International Studies, working draft, May 24, 2006.

Cowell, Alan. "Top General Urges Britain to Leave Iraq." *The New York Times* (October 13, 2006): p. 8.

Dardagan, Hamit, John Sloboda, Kay Williams, and Peter Bagnall. *Iraqi Body Count, A Dossier of Civilian Casualties 2003–2005*. Oxford, UK: Iraqi Body Count Project, 2006. As of October 1, 2007: http://www.iraqbodycount.org

Davison, W. P. *User's Guide to the RAND Interviews in Vietnam*. Santa Monica, Calif.: RAND Corporation, R-1024-ARPA, 1972. Available online at: http://www.rand.org/pubs/reports/R1024/

De Young, Karen, and Colum Lynch. "Bush Abandons Bid to Win U.N. Backing for War." *The Washington Post* (March 18, 2003): p. 16.

Duelfer, Charles, et. al. "Comprehensive Report of the Special Advisor to the DCI [Director of Central Intelligence] on Iraq's WMD [Weapons of Mass Destruction]." Transmittal Message, Washington, D.C., September 30, 2004, p. 9.

Eders, David. "A Million Iraqis Flee War-Torn Country for Haven in Jordan." *Washington Times* (May 27, 2006): p. 6.

Fassihi, Farnaz, Jay Solomon, and Philip Shishkin. "Zarqawi's Death, Completion of Cabinet Raise Hopes in Iraq." *The New York Times* (June 9, 2006): p. 1.

Fay, Major Gen. George R., Investigating Officer. *Investigation of the Abu Ghraib Detention Facility and 205th Military Intelligence Brigade*. Baghdad, 2004.

Finer, Jonathan. "Threat of Shi'ite Militias Now Seen as Iraq's Most Critical Challenge." *The Washington Post* (April 8, 2006).

———. "U.S. Report Cites Progress, Shortfalls in Iraq Rebuilding." *The Washington Post* (May 1, 2006).

Fischer, Hannah. *Iraqi Civilian Death Estimates*. Washington, D.C.: Congressional Research Service, RS22537, 2006.

Fondacaro, Colonel Steve, Chief Joint IED Task Force Forward, Camp Victory, Baghdad. Interview, July 15, 2005.

Franks, Tommy, with Malcolm McConnell. *American Soldier*. New York: HarperCollins, 2004.

Galula, David. "Counterinsurgency Warfare: Theory and Practice." St. Petersburg, Fla.: Hailer Publishing, 1964.

Gambetta, Diego (ed.). *Making Sense of Suicide Bombers*. New York: Oxford University Press, 2005.

Gordon, Michael R. "NATO Moves to Tighten Grip in Afghanistan." *The New York Times* (June 9, 2006).

Gordon, Michael R., and David S. Cloud, "Rumsfeld's Memo on Iraq: Proposed Major Change." *The New York Times* (December 3, 2006).

Gordon, Michael R., and Lt. Gen. (USMC, ret.) Bernard E. Trainor. *Cobra II, The Inside Story of the Invasion and Occupation of Iraq*. New York: Pantheon Books, 2006.

Grossman, Elaine M. "Pace Group to Put Forth Iraq Strategy Alternatives by Mid-December." *Inside the Pentagon* (November 9, 2006).

Hashim, Ahmed S. *Insurgency and Counter-Insurgency in Iraq*. Ithaca, N.Y.: Cornell University Press, 2006.

Hernandez, Nelson. "New Iraqi Power Feeds a Feeble Grid." *The Washington Post* (May 1, 2006): p. A15.

Higgins, Bradford, former Director Joint Strategic Planning and Assessment Branch U.S. Embassy Iraq, currently Assistant Secretary of State and Chief Financial Officer at the U.S. State Department. Interview. Washington, D.C., April 24, 2006.

Hoffman, Bruce. *Does Our Counter-Terrorism Strategy Match the Threat?* Santa Monica, Calif.: RAND Corporation, CT-250, 2005. Available online at: http://www.rand.org/pubs/testimonies/CT250-1/

———. *The Use of the Internet by Islamic Extremists*. Santa Monica, Calif.: RAND Corporation, CT-262-1, 2006.

Hornburg, General (USAF) Hal, Commander, Air Combat Command. Interview. Norfolk, Va., June 3, 2004.

Hosmer, Stephen T. *Why Milosevic Decided to Settle When He Did*. Santa Monica, Calif.: RAND Corporation, MR-1351-AF, 2001. Available online at: http://www.rand.org/pubs/monograph_reports/MR1351/

Hosmer, Steve, and Olya Oliker, *Combating Protracted Counter-Insurgency in Today's Iraq*. Santa Monica, Calif.: RAND Corporation, 2004. Government publication; not releasable to the general public.

Independent Panel on the Safety and Security of the United Nations in Iraq. *Report of Independent Panel on the Safety and Security of the United Nations in Iraq*. New York: United Nations, 2003.

International Crisis Group. *In Their Own Words: Reading the Iraqi Insurgency*. Middle East Report No. 50, February 15, 2006. As of October 1, 2007: http://www.crisisgroup.org/home/index.cfm?id=3953

Jihad, Issam. Statement, reported by Kuwait News Agency, February 18, 2006, quoted by Onur Ozlu, *Iraqi Economic Reconstruction and Development*. Washington, D.C.: Center for Strategic and International Studies, 2006.

Jones, Seth G., et al. *Establishing Law and Order After Conflict*. Santa Monica, Calif.: RAND Corporation, MG-374-RC, 2005. Available online at: http://www.rand.org/pubs/monographs/MG374/

Kalic, Sean N. *Combating a Modern Hydra: Al Qaeda and the Global War on Terrorism*. Fort Leavenworth, Kan.: Combat Studies Institute Press, 2003.

Katzman, Kenneth. *Iraq: Elections, Government, and Constitution*. Washington, D.C.: Congressional Research Service, 2006.

Kellen, Konrad. *RAND Vietnam Motivation and Morale Summary Report*. Santa Monica, Calif.: RAND Corporation, RM-6131-1-ISA, 1970. Available online at: http://www.rand.org/pubs/research_memoranda/RM6131-1/

Khalilzad, Zalmay. "The Next Six Months Will Be Critical." Interview. *Der Spiegel* (June 7, 2006). As of October 1, 2007:
http://service.spiegel.de/cache/international/spiegel/0,1518,419978,00.html

Kilcullen, David. "Twenty-Eight Articles: Fundamentals of Company-Level Counterinsurgency." *Military Review* (2006). As of October 1, 2007: http://usacac.leavenworth.army.mil/CAC/milreview

Knickmeyer, Ellen. "U.S. Will Reinforce Troops In West Iraq." *The Washington Post* (May 30, 2006): p. 1.

Lincoln Group. *Al Anbar Survey 7, 2006 Baseline*. Baghdad: Lincoln Group, May 2006.

Lynch, Kristin F., John G. Drew, Robert S. Tripp, and Charles Robert Roll. *Lessons from Operation Iraqi Freedom*. Santa Monica, Calif.: RAND Corporation, MG-193-AF, 2005. Available online at:
http://www.rand.org/pubs/monographs/MG193/

McDaniel, Colonel (USMC) Edward, Deputy J-3, II Marine Expeditionary Force. Interview, Camp Fallujah, Iraq, August 5, 2005.

McGrath, John J. *Boots on the Ground: Troop Density in Contingency Operations*. Fort Leavenworth, Kan.: Combat Studies Institute Press, 2006.

McIntyre, Captain (Ret) Russ. Interview. Arlington, Va., April 6, 2005.

McMaster, Colonel (USA) H. R. Briefing. Washington, D.C.: RAND Corporation, April 5, 2006.

Michael, George, and Joseph Scolnick. "The Strategic Limits of Suicide Terrorism in Iraq." *Small Wars and Insurgencies*, Volume 17 (June 2006): pp. 113–125.

Mosher, Andy. "U.S.-Backed Operation Targets Shi'ite Slum." *The Washington Post* (August 8, 2006): p. 16.

———. "Baghdad Morgue Tallies 1,815 Bodies in July." *The Washington Post* (August 10, 2006): p. 20.

Moss, Michael, and David Rohde. "Misjudgements Marred U.S. Plans For Iraqi Police." *The New York Times* (May 21, 2006).

Multi-National Force–Iraq. *Counterinsurgency Handbook*. 1st ed. Camp Taji, Iraq: Counterinsurgency Center for Excellence, May 2006.

———. *Provincial Stability Assessment Report*, January 31, 2006.

National Security Council. *National Strategy for Victory in Iraq*. Washington, D.C.: National Security Council, November 2005.

Obaid, Nawaf, *Meeting the Challenge of a Fragmented Iraq: A Saudi Perspective*. Washington, D.C.: Center for Strategic and International Studies, April 6, 2006.

Open Society Institute and the United Nations Foundation. *Iraq in Transition, Post-Conflict Challenges and Opportunities*. New York: Open Society Institute, 2004.

Ozlu, Onur. *Iraqi Economic Reconstruction and Development*. Washington, D.C.: Center for Strategic and International Studies, 2006, p. 30.

Packer, George. "The Lesson of Tal Afar." *The New Yorker* (April 10, 2006).

Pollack, Kenneth M., et al. *A Switch in Time, A New Strategy for America in Iraq*. Washington, D.C.: The Saban Center for Middle East Policy at the Brookings Institution, 2006, pp. 87–89.

Powell, Anita. "They Hate Saddam . . . They Hate Americans." *Stars and Stripes Mideast Edition* (October 23, 2005): p. 4.

Powell, Colin. "Address to the U.N. Security Council, United Nations, New York," February 5, 2003. As of October 1, 2007: http://www.whitehouse.gov/news/releases/2003/02/print/20030205-1.html

"President Bush Announces Major Combat Operations in Iraq Have Ended." Aboard the USS *Abraham Lincoln*, San Diego, Calif., May 1, 2003.

"President Delivers Commencement Address at the United States Military Academy at West Point." New York: Mitchie Stadium, West Point, May 27, 2006. As of October 1, 2007: http://www.whitehouse.gov/news/releases/2006/05/20060527-1.html

"President Says Saddam Must Leave Iraq Within 48 Hours." Washington, D.C.: The Cross Hall, White House, March 17, 2003, p. 1. As of October 1, 2007: http://www.whitehouse.gov/news/releases/2003/03/print/20030317-7.html

Rathmell, Andrew, et al. *Developing Iraq's Security Sector, The Coalition Provisional Authority's Experience*. Santa Monica, Calif.: RAND Corporation, MG-365-OSD, 2005.

Rayburn, Joel. "The Last Exit from Iraq." *Foreign Affairs* (March/April 2006).

Reuters, "Shinseki Repeats Estimate of a Large Postwar Force." *The Washington Post* (March 13, 2003): p. 12.

Sattler, Lt. Gen. (USMC) John F. "Second Battle of Fallujah—Urban Operations in a New Kind of War." Interview with Patricia Slayden Hollis, *Field Artillery Magazine* (March–April 2006): pp. 4–8.

Schlesinger, James R., et al. *Final Report of the Independent Panel to Review Department of Defense Detention Operations*, Washington, D.C., 2004.

Select Committee on Intelligence, United States Senate. *Report on Postwar Findings About Iraq's WMD Programs and Links to Terrorism and How They Compare with Prewar Assessments.* Washington, D.C.: Government Printing Office, 2006.

Shanker, Thom. "U.S. General Says Iraq Could Slide into a Civil War." *The New York Times* (August 4, 2006).

Sharp, Jeremy M., et al. *Post-War Iraq: A Table and Chronology of Foreign Contributions.* Washington, D.C.: Congressional Research Service, 2005.

Slocombe, Walter. Interview, Frontline, Public Broadcasting System, August 17, 2004.

Strasser, Steven (ed.). *The Abu Ghraib Investigations*. New York: Public Affairs, 2004.

Telhami, Shibley. "What Arab Public Opinion Thinks of U.S. Policy." Transcript of Proceedings, Washington, D.C.: The Saban Center for Middle East Policy, The Brookings Institution, 2005.

Tunnell, Lt. Col. (USA), Harry D. III. *Red Devils, Tactical Perspectives from Iraq.* Fort Leavenworth, Kan.: Combat Studies Institute Press, 2004.

Tyson, Ann Scott. "U.S. Studied Bremer's '04 Bid for More Troops." *The Washington Post* (January 13, 2006): p. A17.

United Nations Assistance Mission for Iraq (UNAMI). *Human Rights Report, 1 September–31 October 2006*. New York: United Nations, 2006.

United Nations Security Council, Resolution 1483 (2003), S/RES/1483 (2003), adopted on May 22, 2003.

U.S. Agency for International Development (USAID). "Assistance for Iraq, Electricity." Washington, D.C.: USAID, 2006. As of October 1, 2007: http://www1.usaid.gov/iraq/accomplishments/electricity.html

U.S. Department of Defense. *Directive Number 3000.05, Subject: Department of Defense Capabilities for Stability Operations.* Washington, D.C.: Department of Defense, USD (P), 2006.

———. *Report to Congress, Submitted Pursuant to U.S. Policy in Iraq Act, Section 1227 of the National Defense Authorization Act for Fiscal Year 2006 (PL 109-163).* Washington, D.C.: Government Printing Office, April 6, 2006.

———. *Report to Congress in Accordance with the Department of Defense Appropriations Act 2006 (Section 9010): Measuring Stability and Security in Iraq.* Washington, D.C.: Government Printing Office, August 2006.

U.S. Department of State. *Iraq Weekly Status Report.* Washington, D.C.: U.S. Department of State, Bureau of Near Eastern Affairs, April 19, 2006.

U.S. Government Accountability Office. *Rebuilding Iraq, Stabilization, Reconstruction, and Financing Challenges.* Washington, D.C.: Government Accountability Office, GAO-06428T, 2006.

Weaver, Mary Anne. "Inventing Al-Zarqawi." *The Atlantic*, Vol. 297, No. 6 (July/August 2006): pp. 87–100.

Wilson, Clay. *"Improvised Explosive Devices in Iraq: Effects and Countermeasures.* Washington, D.C.: Congressional Research Service, 2005, p. 1.

Wilson, Colonel GI. Interview. Marine Counter-Terrorism Protection Task Force, Camp Fallujah, August, 5, 2005.

Wimbish, Colonel Calvin. Chief MNF-I, Ministry of Defense, Intelligence Transition Team. Interview. U.S. Embassy, Baghdad, October 10, 2005.

Wiser, Major Conrad. Interview. U.S. Embassy Public Affairs Military Information Support Team (PA-MIST), Baghdad, Iraq, August 12, 2004.

Wolfowitz, Paul, Deputy Secretary of Defense. Testimony before the House Defense Appropriations Subcommittee, Washington, D.C., March 27, 2003.

Woods, Kevin M., et al. *Iraqi Perspectives Project, A View of Iraqi Freedom from Saddam's Senior Leadership.* Norfolk, Va.: Joint Center for Operational Analysis and Lessons Learned, U.S. Joint Forces Command, 2006.

Woodward, Bob. *Bush at War.* New York: Simon & Schuster, 2002.

———. *Plan of Attack.* New York: Simon & Schuster, 2004.

Worth, Robert F. "Blast Destroys Shrine in Iraq, Setting Off Sectarian Fury." *The New York Times* (February 22, 2006).

Wright, Robin. "Bush Initiates Iraq Policy Review Separate from Baker Group's." *The Washington Post* (November 15, 2006).